# Techies Who Talk to Plants

# Advance Praise for *Techies Who Talk to Plants*

This book is a first of its kind, chronicling the extraordinary journeys of young innovators–entrepreneurs who dedicated themselves to developing technology-based solutions for farmers' needs and made them into successful business ventures. A very well-written book!

**V.R. Kaundinya**, Director General, Federation of Seed Industry of India (FSII)

The future of agriculture in India is not just about growing crops but also about cultivating and using innovation. When tradition is supplemented by technology, it creates a harvest of opportunities that can transform not only fields but also the future of the farmers and the country.

**Simon-Thorsten Wiebusch**, Vice Chairman, Managing Director and CEO, Bayer CropScience Ltd; Country Divisional Head, Crop Science Division for India, Bangladesh, and Sri Lanka, Bayer

This book is a refreshing take on how innovation and passion can breathe new life into even the most traditional sectors. A must-read for anyone who believes in the power of innovation to change the world and who is interested in the future of agriculture.

**Dr Panabaka Lakshmi**, former Minister of State for Health and Family Welfare, former Minister of State for Textile

As an investor who is deeply engaged in the agri-tech space and with my experience as the CEO of ITC agri-business, the co-founder of The Agri Collaboratory, and the CEO of The Agri Digital Framework (AgDx), I recognize the significance of the entrepreneurs featured in *Techies Who Talk to Plants*. This book brilliantly captures the innovation and resilience driving transformation in Indian agriculture. These pioneers are not just overcoming challenges but are also laying the foundation for sustainable growth and food security – not only in India but globally. Their journeys serve as an inspiration for all of us invested in the future of agriculture and the immense potential it holds.

**Sanjiv Rangrass**, venture partner, Capria Ventures; angel investor and mentor

Innovation in agriculture is not about technology, it is about reimagining the ways in which food is grown, consumed and distributed to create sustainable food systems for the future generations. The startups mentioned in the book are trying to solve complex problems by working with diverse stakeholders in an industry that is facing challenges due to climate change, growing population, and finite natural resources.

**Venkatram Vasantavada**, Managing Director and CEO, SeedWorks International Limited

# TECHIES WHO TALK TO PLANTS

How India's Agri-Tech Visionaries Are Revolutionizing Farming

SHAH M.M.

Foreword by
Harsh Mariwala

BLOOMSBURY
NEW DELHI · LONDON · OXFORD · NEW YORK · SYDNEY

BLOOMSBURY INDIA
Bloomsbury Publishing India Pvt. Ltd
Second Floor, LSC Building No. 4, DDA Complex, Pocket C – 6 & 7,
Vasant Kunj, New Delhi, 110070

BLOOMSBURY, BLOOMSBURY INDIA and the Diana logo
are trademarks of Bloomsbury Publishing Plc

First published in India 2025

Copyright © Shah M.M., 2025
Foreword copyright © Harsh Mariwala, 2025

Shah M.M. has asserted his right under the
Indian Copyright Act to be identified as the Author of this work

All rights reserved. No part of this publication may be: i) reproduced or transmitted in any form, electronic or mechanical, including photocopying, recording or by means of any information storage or retrieval system without prior permission in writing from the publishers; or ii) used or reproduced in any way for the training, development or operation of artificial intelligence (AI) technologies, including generative AI technologies. The rights holders expressly reserve this publication from the text and data mining exception as per Article 4(3) of the Digital Single Market Directive (EU) 2019/790

ISBN: PB: 978-93-6131-261-8; eBook: 978-93-6131-510-7
2 4 6 8 10 9 7 5 3 1

Typeset in Minion by Manipal Technologies Limited
Printed and bound in India by Gopsons Papers Pvt. Ltd., Noida

To find out more about our authors and books visit www.bloomsbury.com
and sign up for our newsletters

# Contents

*Foreword* by Harsh Mariwala ........ ix

Introduction ........ 1
1. The Destined Disruptor: Sachin Nandwana, BigHaat ........ 19
2. The Intuitive Innovator: Prateep Basu, SatSure ........ 51
3. The Tinkerer Turned Titan: Karthik Jayaraman, WayCool ........ 78
4. The Accidental Aquaman: Rajamanohar, Aquaconnect ........ 112
5. The Achrekar of Agri-Tech: Hemendra Mathur, Investor and Mentor ........ 140
6. The Yolk Reformer: Abhishek Negi, Eggoz ........ 155
7. The Purpose-Driven Founder: Nidhi Pant, S4S Technologies ........ 183
8. The Unlikely Champion of Innovation: Shridhar Mehta, Prompt Dairy Tech ........ 205
9. The Barefoot Entrepreneur: Shashank Kumar, DeHaat ........ 225
10. Sowing the Seeds of Change: Lessons from the Field ........ 250

*Acknowledgements* ........ 255
*Index* ........ 257
*About the Author* ........ 259

# Foreword

THE INDIAN AGRICULTURAL SECTOR has seen significant growth since the country's independence in 1947, ensuring a movement from food deficit to homegrown food security.

Agriculture yield has been traditionally dependent on seasonal and climatic factors. In the coming decades, with climate change, the role of science and innovation in agriculture will be essential to mitigate any potential reversal of the growth that we have been able to achieve.

With rising incomes, shifting consumer preferences, and increased awareness about the quality of products, the agricultural sector needs to evolve and adapt to meet the demands of a changing world and feed billions of people. In India, where agriculture supports over half of the population but struggles with issues such as fragmented land holdings, water scarcity, and outdated technology, innovation can be a game changer and enabler of large-scale production.

Moving from traditional practices and adopting new technologies will involve cooperation and assistance from the government as well as other institutions that can provide farmers with support and back them to take risks. I have always viewed Marico with a vision of not only building a business but also giving back to society.

As a consumer brand with deep roots in India's agrarian ecosystem, Marico's CSR (Corporate Social Responsibility) initiatives, like the Parachute Kalpavriksha Foundation, support 65,000 farmers in Tamil Nadu,

Kerala, Andhra Pradesh, and Karnataka. A team of over 100 agronomists regularly visits the farmers and educates them on scientific farming practices and helps them increase their productivity. Through the Marico Innovation Foundation, we identify and provide bespoke mentorship and capacity-building support to innovative agriculture startups to scale up.

With many such interventions emerging in the country, I am delighted to observe the rising number of agriculture startups. As per a report released by the Federation of All India Farmer Associations, there were less than 50 startups in agriculture and allied sectors prior to 2014–15; that number has crossed 7,000 in the last ten years due to a conducive business environment and government support.[1]

Unlike other industries, where developing proofs of concept can be accomplished in a few months, agricultural entrepreneurs face the unique challenge of longer cropping cycles extending for more than a year, which significantly extends the timeline for testing and validation. For this reason, along with other challenges of uncertainty, low margins, and deeply entrenched traditional practices, these entrepreneurs need a vision that goes beyond creating profits. It takes not just an entrepreneurial spirit but a profound sense of purpose to persevere in the face of such obstacles.

What I find particularly inspiring is how these entrepreneurs are not only building scalable, profitable businesses but also contributing to India's larger goals of food security and rural prosperity. By creating solutions that address real-world challenges, they are helping to ensure that India's agricultural sector can feed its growing population while improving the livelihoods of millions of farmers. To me this intersection of profit and purpose is the future of business.

---

[1] Press Trust of India, 'Number of Agri Startups Jumps Multifold to 7,000 in Last 9 Years: Report', *Business Standard*, 15 May 2024.

One of the remarkable aspects of agri-tech startups is their ability to bridge the rural–urban divide. Many entrepreneurs come from urban, tech-focused backgrounds but are finding innovative ways to connect with rural communities. By immersing themselves in the challenges farmers face and designing solutions based on real-world needs, these startups are gaining the trust of rural farmers. Through technology, they are helping farmers integrate into the larger economy while maintaining the cultural and social fabric of rural life.

Innovation and startups are at the forefront of transforming agriculture, making it more efficient, sustainable, and profitable. By combining technology with a deep understanding of the challenges faced by farmers, these ventures are not only addressing existing problems but also creating new opportunities for growth and development.

As more entrepreneurs take up the mantle of building the future of agriculture, the sector is poised for a period of unprecedented innovation and transformation, thereby helping improve food security, enhancing rural livelihoods, and contributing to the nation's economic progress.

This book is not just a collection of stories; it is a roadmap for anyone looking to build a purpose-driven business. The entrepreneurs featured here serve as role models for the next generation of innovators, demonstrating that it is possible to build successful, scalable businesses while staying true to a mission that benefits society.

As I reflect on the journeys chronicled in this book, I am filled with a deep sense of admiration for the entrepreneurs who have chosen to work in the agriculture industry. They are not only transforming the sector but also showing us what is possible when purpose, passion, and innovation come together. I am confident that their stories will inspire countless others to take up the mantle and continue the work of transforming Indian agriculture.

**Harsh Mariwala**
Chairman, Marico Ltd

# Introduction

## The Evolution of Agribusiness in India: From Family Enterprises to Tech Startups

India's agricultural landscape reflects centuries of tradition, innovation, and resilience. As the backbone of the nation's economy, agriculture has been both a source of livelihood for millions and the cornerstone of food security for what is now the world's most populous country. The sector contributes approximately 20 per cent to India's GDP and employs nearly half of the country's workforce.

The vast expanse of India's agricultural land, spanning diverse climatic zones, from the Himalayan foothills to the coastal plains, supports a wide variety of crops. This diversity has given rise to a complex ecosystem of businesses that support, enhance, and derive value from agricultural activities. From small family-owned enterprises that have existed for generations to modern tech startups disrupting traditional practices, the story of agribusiness in India is one of constant evolution and adaptation.

As we delve into this narrative, we'll touch briefly on how the landscape of agribusiness has evolved over time, influenced by various factors. This overview will include a quick look at the shift from traditional family-owned businesses to the rise of corporate entities and the recent emergence of agri-tech startups in India.

### *Traditional family-owned agribusinesses*

The roots of agribusiness in India run deep, intertwined with the country's agrarian history. For centuries, family-owned enterprises have been the bedrock of the agricultural sector, passing down knowledge, skills, and business acumen from one generation to the next.

These family-owned businesses have traditionally dominated various segments of the agricultural value chain. Let's explore some of the key types of these businesses:

1. **Seed and fertilizer companies:** Many of India's prominent seed and fertilizer businesses started as small family operations. Companies like Mahyco (Maharashtra Hybrid Seeds Company), which began as a modest venture by Dr Badrinarayan Barwale in 1964, have grown to become leaders in the seed industry. These businesses played a crucial role in the Green Revolution. This period saw the introduction of high-yield varieties of crops and modern farming techniques that dramatically increase agricultural productivity. The Green Revolution in India gained momentum in the late 1960s and reached its peak in the early to mid-1970s, transforming the country from a food-deficient nation to one capable of achieving self-sufficiency in food grain production.

2. **Farm equipment manufacturers:** The mechanization of Indian agriculture owes much to family-owned businesses that started manufacturing farm equipment. Firms like the Sonalika Group, which began as a small tractor assembly unit in the 1960s, have grown into major players in the agricultural machinery sector.

3. **Food processing units:** From small-scale mills to larger processing plants, family-owned food processing units have been integral to India's agribusiness landscape. These businesses, often started to cater to local needs, have sometimes grown into national brands. For instance, Haldiram's began as a small shop in Bikaner and is now a global Indian snack food company.

4. **Agricultural trading firms:** Family-owned trading businesses have long facilitated the movement of agricultural produce from farms to markets. Many of these firms started as small operations in local mandis (markets) and expanded over time to become significant players in agricultural commodities trading.

The characteristics of these family-owned businesses often include strong local connections, a deep understanding of regional agricultural practices, and a high degree of trust within their communities. Many have built their reputations over decades, leveraging personal relationships and word-of-mouth recommendations to grow their customer base.

However, these businesses also face unique challenges. The informal nature of many family enterprises can lead to difficulties in accessing credit and scaling operations. Succession planning is another critical issue, with the next generation sometimes reluctant to continue in the family business and preferring instead to pursue careers in other sectors.

Despite these challenges, many family-owned agribusinesses have shown remarkable resilience and adaptability. Some have successfully transitioned to professional management structures while retaining family ownership, allowing them to compete effectively in an increasingly globalized market.

*Corporate agriculture*

As India's economy began to liberalize in the 1990s, the agricultural sector saw the entry of large corporations, both domestic and international. This marked a significant shift in the agribusiness landscape, introducing new models of operation and scales of production previously unseen in India.

The entry of corporate players brought with it substantial capital investment, advanced technologies, and modern management practices. Companies like ITC, Godrej Agrovet, and Mahindra & Mahindra's agribusiness division began to make significant inroads into various aspects of agriculture, from input supply to food processing and distribution.

One of the most notable developments in this era was the rise of contract farming. Under this model, companies enter into agreements with farmers to produce specific crops under pre-agreed terms. This approach offered several advantages:

1. For farmers, it provided a guaranteed market for their produce and often included support in the form of inputs, technology, and advisory services.
2. For companies, it ensured a steady supply of quality produce tailored to their specifications.

Pepsi's tomato farming initiative in Punjab is often cited as an early success story of contract farming in India. The company worked with farmers to grow tomato varieties suitable for processing, providing them with seedlings, agronomic support, and a buy-back guarantee.

Vertical integration became another hallmark of corporate involvement in agriculture. Companies began to control multiple stages of the value chain, from farm to fork. For instance, Godrej Agrovet not only produces animal feed but also operates in the poultry and dairy sectors, making for an integrated business model.

The impact of corporate agriculture on traditional farming practices and markets has been significant and multifaceted. On one hand, it has introduced efficiencies, improved quality standards, and opened new market opportunities for farmers. The infusion of technology and modern practices has helped increase productivity in many areas.

On the other hand, the growing corporate presence in agriculture has raised concerns about the concentration of market power and its potential to marginalize small farmers. Critics of contract farming argue that while it is beneficial in many ways, it can also lead to a loss of autonomy for farmers and potentially trap them in unfavourable long-term agreements.

Moreover, the entry of large players has put pressure on traditional family-owned businesses, particularly in areas like seed production and distribution, where economies of scale play a crucial role. Some family businesses have adapted by becoming part of the supply chain for larger corporations, while others have struggled to compete.

## Cooperative movements in agriculture

As corporate agriculture gained ground in India, the cooperative movement emerged as a counterbalance, aiming to empower small and marginal farmers through collective action. Agricultural cooperatives in India have a long history, dating back to the early 20th century, but they gained prominence in the post-Independence era.

Cooperatives play a crucial role in aggregating the resources and produce of small farmers, giving them greater bargaining power in the market. They operate across various segments of the agricultural value chain. They consist of:
1. Credit cooperatives, providing financial services to farmers
2. Marketing cooperatives, helping farmers sell their produce collectively
3. Input supply cooperatives, offering seeds, fertilizers, and other inputs at competitive prices
4. Processing cooperatives, adding value to farm produce

One of the most celebrated success stories in the cooperative movement is Amul, run by the Gujarat Cooperative Milk Marketing Federation. Started in 1946, Amul revolutionized milk production and distribution in India, empowering millions of small dairy farmers. The 'Anand pattern' of cooperatives, pioneered by Amul, has been replicated across India and in other developing countries.

Another notable example is the Indian Farmers Fertilizer Cooperative Limited (IFFCO), one of the largest fertilizer manufacturers in the world. IFFCO has played a significant role in ensuring the availability of quality fertilizers to Indian farmers at affordable prices.

However, the cooperative movement in Indian agriculture has also faced numerous challenges. These include:

1. Political interference in the management of cooperatives
2. Lack of professional management and modern business practices
3. Difficulty in adapting to changing market conditions
4. Financial constraints and dependence on government support

Despite these challenges, agricultural cooperatives continue to play a vital role in supporting small farmers and maintaining a balance in the agribusiness ecosystem in the country. Many cooperatives are now adopting modern technologies and business practices to remain competitive in the evolving market landscape.

## *Government initiatives and public-sector enterprises*

The Indian government has always played a significant role in shaping the country's agricultural sector and, by extension, its agribusiness landscape. Through various initiatives, policies, and public sector enterprises, the government has sought to support farmers, ensure food security, and promote agricultural development.

One of the most prominent public-sector entities in Indian agriculture is the Food Corporation of India (FCI). Established in 1965, the FCI plays a crucial role in implementing the government's food policies. Its primary functions include:
1. Procurement of food grains at Minimum Support Prices (MSP)
2. Distribution of food grains through the Public Distribution System (PDS)
3. Maintenance of buffer stocks to ensure food security

Other important public-sector enterprises in the agricultural sector include:
1. **National Seeds Corporation (NSC)**: Responsible for producing and distributing high-quality seeds
2. **National Agricultural Cooperative Marketing Federation of India (NAFED)**: Focused on promoting cooperative marketing of agricultural produce
3. **National Bank for Agriculture and Rural Development (NABARD)**: Provider of credit for the promotion of agriculture and rural development

In recent years, the government has also been actively promoting public-private partnerships (PPPs) in agribusiness. These partnerships aim to leverage private-sector expertise and resources while addressing public-sector goals. Examples of such initiatives include:
1. **The National Agriculture Market (eNAM)**: An electronic trading portal that connects existing APMC (Agricultural Produce Market Committee) mandis to create a unified national market for agricultural commodities
2. **Kisan Rail**: A partnership between the Ministry of Railways and the Ministry of Agriculture and Farmers' Welfare to facilitate fast transportation of agricultural produce

3. **Mega Food Parks Scheme**: Aimed at creating modern food processing infrastructure with a well-established supply chain

These government initiatives and public-sector enterprises have played a crucial role in supporting both traditional family-owned businesses and emerging agri-tech startups. They have helped create an enabling environment for agribusiness, providing infrastructure, market access, and financial support.

However, the effectiveness of government interventions in agriculture has been a subject of debate. Critics argue that some policies, while well-intentioned, have led to market distortions and unsustainable practices. The challenge lies in striking a balance between necessary support and market-driven growth.

## *The rise of agri-tech startups*

In recent years, India has witnessed a surge in agri-tech startups, marking a new chapter in the evolution of its agribusiness landscape. This phenomenon is driven by a confluence of factors:
1. Increased internet and smartphone penetration in rural areas
2. The government's push for initiatives like Digital India and Startup India
3. Growing awareness of the need for sustainable and efficient agricultural practices
4. Availability of venture capital funding for agriculture-focused technology companies

Unlike traditional family-owned businesses that often rely on inherited knowledge and established practices, agri-tech startups are leveraging cutting-edge technologies to address long-standing challenges in Indian agriculture.

Many of these startups are founded by young entrepreneurs, often graduates from prestigious institutions like the Indian Institutes of Technology (IITs) or Indian Institutes of Management (IIMs).

These new-age entrepreneurs bring a fresh perspective to agribusiness. They combine technological expertise with a deep understanding of agricultural challenges, often gained through extensive field research and collaboration with farmers. This approach stands in contrast to that of traditional family-owned businesses, which typically evolve their practices incrementally over generations.

The key areas of innovation in the agri-tech space include:

1. **Precision farming:** Startups are using the Internet of Things (IoT), sensors, drones, and satellite imagery to provide farmers with real-time data on soil health, weather conditions, and crop growth. This enables more precise application of inputs and better crop management.
2. **Supply chain optimization:** Some companies are revolutionizing the agricultural supply chain by connecting farmers directly with retailers and consumers. These platforms use data analytics and logistics optimization to reduce wastage and improve farm incomes.
3. **Farm management software:** Innovative firms are developing comprehensive farm management solutions that help farmers digitize their operations, from planning and budgeting to harvesting and sales.
4. **Agri-fintech solutions:** Several startups are addressing the perennial problem of access to credit for small farmers. They use alternative data sources and innovative risk assessment models to provide faster and more accessible financial services to the agricultural sector.

## The changing face of Indian agribusiness

The emergence of agri-tech startups alongside traditional family-owned businesses and corporate players has created a diverse and dynamic agribusiness ecosystem in India. This coexistence of the old and new is reshaping the sector in several ways:

1. **Technology adoption:** The innovations introduced by agri-tech startups are gradually being adopted by traditional businesses and corporate players, leading to a broader technological transformation across the sector.
2. **Changing business models:** The direct-to-farmer and direct-to-consumer models popularized by many agri-tech startups are influencing how traditional businesses operate, pushing them to become more efficient and customer-centric.
3. **Focus on sustainability:** Many agri-tech startups emphasize sustainable practices, encouraging the broader industry to pay more attention to environmental concerns.
4. **Skill development:** The growing need for tech-savvy professionals in agriculture is leading to new opportunities for rural youth and changing the skill profile of the agricultural workforce.

However, this transition is not without challenges:

1. **The digital divide:** The reliance on technology by agri-tech startups can potentially exclude farmers who lack access to smartphones or the internet, particularly in remote areas.
2. **Trust building:** Many farmers are hesitant to adopt new technologies or change long-standing practices. Agri-tech startups face the challenge of building trust and demonstrating the tangible benefits of their solutions.

3. **Regulatory environment:** The regulatory framework in India is still catching up with the pace of innovation in agri-tech. Issues around data privacy, drone usage, and digital payments in rural areas need to be addressed.
4. **Scaling up:** While many agri-tech startups offer innovative solutions, scaling them up to reach millions of small and marginal farmers across India's diverse agro-climatic zones remains a significant challenge.

## Future outlook

As we look at the future of agribusiness in India, several trends and possibilities emerge:

1. **Collaboration between the traditional and modern:** There's a growing potential for collaboration between traditional family-owned businesses and agri-tech startups. The deep market knowledge and established networks of traditional businesses combined with the technological innovations of startups could create powerful synergies.
2. **Data-driven agriculture:** The role of data in agriculture is set to grow exponentially. From predictive analytics for crop planning to precision agricultural techniques, data will drive decision-making at every level of the agricultural value chain.
3. **Sustainable and climate-resilient farming:** With climate change posing significant challenges to agriculture, technologies that promote sustainable practices and enhance climate resilience will become increasingly important.
4. **Biotechnology and genetic engineering:** Advancements in biotechnology could lead to the development of crops that are more resistant to pests, diseases, and environmental stresses.

5. **Vertical and urban farming:** As urbanization continues, technologies enabling vertical and urban farming could gain prominence, bringing food production closer to consumption centres.
6. **Blockchain in agriculture:** Blockchain technology could revolutionize agricultural supply chains, enhancing traceability and transparency from farm to fork.
7. **Artificial intelligence and robotics:** AI-powered decision-support systems and agricultural robots could become more common, addressing labour shortages and improving efficiency.

Technology has the potential to address long-standing agricultural challenges, such as low productivity, water scarcity, and market inefficiencies. However, the key to realizing its potential lies in ensuring that technological advancements are inclusive and accessible to all farmers, regardless of the technologies' scale of operation or the farmers' level of education.

As India strives to double farmer incomes and ensure food security for its growing population, the agribusiness sector will play a crucial role. The future success of Indian agriculture will depend on how well it can balance tradition with innovation, leverage technology while preserving ecological balance, and create an ecosystem where small farmers, family-owned businesses, large corporations, and tech startups can all thrive together.

## Pioneering Change: The Agri-Tech Entrepreneurs

The emergence of agri-tech startups, as discussed earlier, marks a significant evolution in India's agricultural landscape. These innovative ventures are not just introducing new technologies; they're fundamentally reimagining how agricultural businesses operate. By leveraging data and digital platforms, they're creating more transparent, efficient, and farmer-centric systems.

## The unique landscape of Indian agri-tech

The Indian agri-tech ecosystem is unlike any other in the world. It operates within a complex web of cultural traditions, diverse agro-climatic zones, and socio-economic realities that are uniquely Indian. Understanding how entrepreneurs navigate this landscape provides valuable insights not just for India but for the global agricultural community as well.

While much has been written about tech startups in urban India, the rural innovation story remains largely untold. This book aims to bridge that gap, showcasing how entrepreneurship and technology are transforming lives in India's hinterlands. It's a story of connecting India with Bharat — the rural heart of the nation, which is often overlooked in discussions on India's economic progress.

The popular narrative of Indian startups often revolves around names like Redbus, Flipkart, and Paytm. But these success stories, while impressive, represent only a fraction of India's potential. The true test of India's entrepreneurial spirit lies in addressing the challenges faced by Bharat, and agri-tech will be at the forefront of this mission.

## Entrepreneurship without a roadmap

What sets agri-tech entrepreneurs apart is the sheer originality of their journey. Unlike Indian entrepreneurs in many other sectors where they could look to Silicon Valley or other global hubs for inspiration and models, agri-tech entrepreneurs in India are writing their own scripts. There's no precedent to follow, no established playbook to consult.

These pioneers are creating original solutions tailored to the unique challenges of Indian agriculture, and in doing so, they're creating models that are now being emulated around the world.

This lack of precedent extends to the funding ecosystem too. While venture capital has flowed freely into many sectors of the Indian startup economy, agri-tech has long been overlooked. Even a few years ago, finding investors willing to back agri-tech ventures was a Herculean task. Today, while the situation has improved, with a handful of VCs showing interest, it still remains a challenging landscape for fundraising.

## *The real face of agri-tech entrepreneurship*

Building an agri-tech business is not for the faint-hearted. It demands a level of commitment and grit that goes far beyond what's required in many other sectors. For these entrepreneurs, success isn't built in air-conditioned offices or gleaming co-working spaces; it's forged in the fields, under the scorching sun, alongside the very farmers they aim to serve.

This book pulls back the curtain on the real, often gritty world of agri-tech entrepreneurship. It's a world where an IIT graduate might trade in his blazer for muddy boots, where a former globe-trotting executive might find herself sleeping in a village hut or a school building because there simply aren't any hotels available in the locality she is working in. It's about spending days, months, and years in remote villages and farms, not just visiting but becoming a part of the community.

Success in this field requires more than just technical knowledge or business acumen. It calls for the ability to connect with farmers on their own turf, to earn their trust, and to prove that you are truly one of them. It means negotiating with tough middlemen, sometimes confronting local power structures, and always being prepared for the unexpected.

These entrepreneurs face challenges that would be unimaginable in most other sectors. They have to navigate complex policy landscapes, deal with the vagaries of weather, and confront deeply entrenched systems that often resist change. Yet, they persist, driven by their vision of what Indian agriculture could become.

## The broader impact

The importance of agri-tech entrepreneurship extends far beyond the boundaries of India. In an era of climate change and growing food insecurity, the innovations emerging from Indian farms and agri-tech startups have global relevance. From AI-powered crop monitoring to blockchain-enabled supply chains, the solutions being developed in India have the potential to transform agriculture worldwide.

Moreover, the agri-tech sector is tackling some of the most complex problems facing humanity today. Food security, sustainable resource management, and rural economic development are all intricately linked to the success of agricultural innovation. By focusing on these challenges, agri-tech entrepreneurs are not just building businesses; they're contributing to solutions for some of the most pressing issues of our time.

## A sector in constant flux

One of the unique aspects of the agri-tech sector is its inherent volatility. Unlike many other industries, where success can be more easily quantified and sustained, agri-tech businesses are subject to a wide range of external factors that can dramatically impact their trajectory.

This book doesn't shy away from this reality. While it celebrates the efforts and innovations of agri-tech entrepreneurs, it also presents a realistic picture of the challenges they face. A company might achieve remarkable growth, reaching revenues of thousands of crores of rupees, only to face setbacks resulting from policy changes, weather events, or market fluctuations. The impact of these external factors is often far more pronounced in agriculture than in other sectors.

As a result, the definition of success in agri-tech is nuanced. It's not just about financial metrics or growth rates; it's about resilience, adaptability, and the ability to create lasting positive change in the lives of farmers and rural communities. The entrepreneurs featured in this book are pioneers, and like all pioneers, they've taken their share of hits. Their stories serve not just as an inspiration but as valuable lessons for the next generation of agri-tech innovators, who may have a smoother path thanks to the groundwork laid by these trailblazers.

## *Lessons beyond agri-tech*

While the focus of this book is agri-tech, the lessons it contains have broader applicability. The strategies these entrepreneurs have used to build businesses in one of the most challenging sectors in the country offer insights that can be valuable across industries. In fact, after exploring hundreds of business stories from various sectors, I can confidently say that the lessons that can be learned from these agri-tech startups are unique and profoundly instructive.

These stories offer a master class in resilience, innovation, and the art of building businesses that create genuine social impact. They demonstrate how complex stakeholder ecosystems can be navigated, how innovation can happen with limited resources, and how value can be created in markets that many consider too difficult to enter.

## The road ahead

Despite the progress made by the entrepreneurs featured in this book, the potential of the Indian agri-tech sector remains largely untapped. The cumulative market share occupied by these businesses is still marginal compared with the vast potential of Indian agriculture. This represents an enormous opportunity for new entrants to make their mark and build lasting, impactful businesses.

As Hemendra Mathur, a veteran in the agri-tech space, points out, 'India has the potential to build another revolution like the IT revolution of the 1990s in Agriculture.' This book aims to catalyse that revolution by inspiring more entrepreneurs to enter the field, encouraging more venture capitalists to invest in agri-tech, and advocating for supportive government interventions.

Moreover, this book serves as a call to action for consumers. In an age where awareness of the source of our food and its environmental impact is increasingly crucial, these stories highlight the importance of supporting innovative, sustainable agricultural practices.

## A celebration of effort and vision

Ultimately, this book is not about celebrating individual successes or lionizing particular enterprises. It's about recognizing and honouring the extraordinary efforts of those who have chosen to dedicate their lives and careers to one of the most challenging yet vital sectors of the economy.

These entrepreneurs are the unsung heroes of India's economic story. They're bridging the gap between India and Bharat, bringing technological innovation to the heartland, and working tirelessly to ensure food security for millions. Their journey is ongoing, their challenges ever-evolving, but their commitment remains unwavering.

As you delve into the pages that follow, you'll encounter stories of triumph and setback, innovation and perseverance. You'll gain insights into how businesses are built from the ground up in a sector where there are no easy wins. Most importantly, you'll come to understand why agri-tech entrepreneurship is not just a business opportunity, but a calling – one that has the potential to reshape the future of India and the world.

Whether you're an aspiring entrepreneur, an investor looking for the next big opportunity, a policymaker seeking to understand the rural innovation landscape, or simply a curious reader interested in the future of food and agriculture in the country, this book offers you valuable insights and inspiration. It's an invitation to be part of a revolution – one that's growing in the fields of India and reaching for the stars.

# 1

## The Destined Disruptor:
## Sachin Nandwana, BigHaat

*IN THE DIVERSE LANDSCAPE of entrepreneurship, we encounter a myriad of origin stories: the accidental innovators, the reluctant leaders, the opportunistic visionaries. Yet, from among these varied paths emerges a rarer breed – the 'destined' entrepreneur. Sachin Nandwana embodies this archetype, with his journey from a farm boy in Rajasthan to the co-founder of BigHaat unfolding with a sense of inexorable purpose. While many harbour dreams of entrepreneurial glory, Sachin refused to entrust his ambitions to fate. From his formative years, he displayed an acute awareness of the skills he needed to acquire, the experiences he had to amass, and the knowledge gaps he needed to bridge to get to where he wanted. With the methodical patience of a seasoned farmer, he cultivated his abilities, navigating corporate roles and educational pursuits with an unwavering focus on his ultimate goal. In the unpredictable terrain of agri-tech, where luck often plays a critical role, Sachin's story serves as a powerful reminder that true success stems not just from ambition but also from meticulous preparation and steadfast determination. His rise proves that while entrepreneurship may be a calling for many, for some, like Sachin, it is an inevitability shaped by choice, not chance.*

Sachin Nandwana's entrepreneurial journey is deeply rooted in his early experiences growing up in a farming family in Rajasthan. As a child, he spent summer vacations working alongside his maternal grandmother in the fields of his ancestral village, gaining first-hand knowledge of the challenges faced by farmers and the intricacies of agriculture. The Nandwana household valued entrepreneurship and self-reliance as core principles. Sachin's grandfather and extended family members were all either farmers or ran their own businesses, never working for others. His father made the decision to seek employment in a public-sector unit and move the family to the town of Kota but still encouraged his son to pursue his own entrepreneurial path. This family environment instilled in Sachin the ethos of problem-solving and value creation from a young age.

Even in his early years, Sachin's business acumen was evident. Throughout school, he engaged in various small-scale ventures that honed his skills. He would purchase maps at 5 paise and sell them for 25 paise, discovering the power of arbitrage and mark-up pricing. He also experimented with buying second-hand books from waste-paper markets, refurbishing them, and reselling them to used bookstores at a significant profit.

However, as he attempted to scale up these ventures, he began to understand the importance of business model scalability and sustainability. He realized that many of his early ideas, while profitable on a small scale, lacked the long-term growth potential and market depth required to evolve into large, enduring enterprises.

After completing his 12th standard, Sachin found himself at a crossroads. His father, supportive of his inclinations, encouraged him to start a small shop. But Sachin had bigger aspirations. He recognized that whether in running a small retail outlet or a larger business, the demands on his time and effort would be similar. He wanted to build a truly scalable venture, and he knew he needed to acquire more knowledge, skills, and experience to do so. Moreover, Sachin was driven by a deep desire to create meaningful value for customers. He wasn't content with merely starting a business; he wanted to solve real problems and make a positive impact. This quest for knowledge, skills, and purposeful entrepreneurship led him to pursue an engineering degree at Mandsaur Institute of Technology, Indore.

During his undergraduate years, Sachin's entrepreneurial spirit thrived alongside his academic pursuits. He set up a personal lab to experiment with electronic designs and prototypes, focusing on real-world problems. One such issue was the frequent power outages plaguing households in the early 2000s, which inspired him to innovate in the emergency light space.

Driven by his belief in value addition, Sachin sought to enhance the utility, affordability, and aesthetic appeal of emergency lights. He explored cost-reduction strategies, such as using lightweight materials and removable batteries, and added functionality by incorporating mobile phone charging ports. He also envisioned lights that could double as decorative pieces, aiming to create a product that was both practical and aesthetically pleasing.

Despite his innovative ideas, Sachin realized that the emergency light business had limited growth potential because of eventual market saturation. This was a recurring theme in his early entrepreneurial experiments. While he excelled at ideating and prototyping, the pathway to building an enduring, high-impact business remained elusive.

Through these experiences, Sachin recognized that truly successful ventures needed three key elements: a large, addressable market, a clear competitive advantage, and the potential for continuous innovation and expansion. This insight would prove crucial in shaping his future endeavours.

As he approached the end of his undergraduate studies, Sachin found himself at another inflection point. While his engineering education was valuable, he recognized it wasn't sufficient to tackle the complex challenges of building a scalable, impactful business. He knew he needed more exposure to real-world business challenges and industry dynamics before pursuing full-time entrepreneurship. To bridge this gap, he decided to seek professional experience that would deepen his domain expertise, hone his problem-solving skills, and provide insights into successful organizations. With this intent, Sachin joined Instrumentation Limited in Kota as an engineer in 2004, viewing it as an opportunity to gain hands-on technical experience while learning about business operations, customer service, and organizational dynamics.

At Instrumentation Limited, Sachin quickly distinguished himself as a versatile problem solver with a founder's mindset. When assigned to the struggling service department, he saw an opportunity to make a significant impact rather than simply perform his assigned clerical tasks.

Assessing the department's overall performance, he identified several areas for improvement. Sachin noticed that many services were being provided free of charge under product warranties, even when problems were caused by customer mishandling. By implementing a policy to charge for such services, he not only reduced unnecessary costs but also encouraged customers to handle products more carefully. He also introduced delivery fees for serviced products, which customers willingly paid for the added convenience.

His innovative approach extended to streamlining internal processes at the company. Sachin proposed bringing quality control functions and preliminary testing into the service department itself, training service personnel to perform final inspections. This eliminated the need for products to be sent to a separate engineering department, saving time and resources while enabling faster turnaround times for customers.

Observing that the expert team was spending hours analyzing simple issues, he implemented a triage system to quickly identify and address minor repairs. This allowed the team to focus on more complex issues and deliver faster service to customers.

As Sachin's initiatives took hold, the service department's performance improved dramatically. Within months, the quarterly turnover equalled three-quarters of its previous annual revenue, transforming a loss-making operation into a profitable one. This success not only demonstrated his strategic thinking but also reinforced his belief in the power of entrepreneurial thinking within an organizational setting.

The three-year stint at Instrumentation Limited provided Sachin with a solid foundation in business operations and customer service while honing his problem-solving skills and ability to drive change. Encouraged by this experience, he took his next significant step, joining the prestigious Bharat Electronics Limited (BEL).

At BEL, Sachin gained exposure to cutting-edge technologies and industry trends. He noted the growing preference for digital screens, laptops, and miniaturization, astutely recognizing that the future lay in embedded systems and IoT. Eager to position himself at the forefront of this technological revolution, Sachin saw the pursuit of an advanced course in embedded systems at the Centre for Development of Advanced Computing (C-DAC) as the solution.

This decision, however, came with its own set of challenges. Enrolling in the course meant leaving behind the stability and financial security of his job at BEL, a prestigious public-sector undertaking. With a young family to support, including a one-and-a-half-year-old son, the prospect of potentially putting his family's financial well-being at risk weighed heavily on his mind. The economic uncertainty in the wake of the 2008 global financial crisis only added to his dilemma, as leaving a stable job seemed risky, with no guarantee of finding another one after completing the course.

Despite these concerns, Sachin's entrepreneurial spirit pushed him to apply for the C-DAC entrance exam, which he cleared with an impressive sixth rank. However, the weight of his responsibilities and the uncertainty about his future prospects caused him to hesitate, missing the initial admission deadline.

Fate, it seemed, had other plans. During one of his weekly commutes between Kota (where his family stayed) and BEL, Mumbai, Sachin received an unexpected message from C-DAC. The institute had decided to increase its intake of students for the embedded systems course, a rare occurrence, and his name was once again on the list, giving him another chance to pursue his dream. As if by divine intervention, his train to Mumbai was delayed by several hours, providing him ample time to contemplate this unexpected opportunity.

Upon reaching home from the railway station, Sachin rushed to discuss the situation with his wife, who stood by his side and believed in his vision of mastering advanced embedded technology to pioneer future innovations. Together, they explored potential outcomes and worst-case scenarios. During this crucial conversation, a moment of clarity struck. Sachin realized that even in the worst-case scenario, he could return to his hometown and start a small coaching institute teaching embedded systems to students.

Armed with this new-found perspective and his wife's support, Sachin made the bold decision to resign from BEL and enrol in the C-DAC's PG Diploma in Embedded Systems and Design (PG-DESD) programme. He understood that this leap of faith would require sacrifices and challenges, but he felt prepared to face them head-on, driven by his passion for technology and desire to create something meaningful.

During his time at C-DAC, Sachin immersed himself in the world of embedded systems, gaining a deep understanding of the technologies and methodologies that would shape the future of connected devices. As he neared the end of his programme, he began to crystallize plans for establishing an embedded systems coaching institute in his hometown of Kota. However, another unexpected opportunity arose – an interview with Honeywell, a global technology giant. Initially hesitant, Sachin soon realized that this could be more than just a job; it could be a valuable learning experience. He saw it as a chance to gather insights that would prove invaluable for his future entrepreneurial ventures.

The interview with Honeywell went exceptionally well, and Sachin was offered a role in the company. As he began his tenure, he found himself immersed in a dynamic environment, working on cutting-edge projects. Although the job was more software-focused than expected, Sachin embraced the opportunity to learn about modern tech development, including agile methodologies and cross-functional collaboration.

While excelling in his role, he never lost sight of his entrepreneurial ambitions. Sachin dedicated his spare time to ideating and refining potential business concepts, actively seeking out like-minded colleagues for discussions. It was during one such conversation that he met Sateesh Nukala, a kindred entrepreneurial spirit. Despite working in different departments at the same company, they discovered a shared passion for leveraging technology to solve real-world problems and create value for customers. They began collaborating, brainstorming ideas, and exploring potential business opportunities that aligned with their skills and interests.

In 2013, an exciting opportunity presented itself when an Irish firm approached them to develop an electronic product. Sachin and Sateesh, still employed at Honeywell, threw themselves into the project, working on it part-time and often sacrificing their evenings and weekends. They were tasked with cost and value engineering for the Indian market and secured rights for product distribution in South Asia. The duo worked closely with the Irish firm in co-developing the product, refining the technology, and gathering user feedback.

As they began piloting and attempting to sell the product in India, Sachin and Sateesh delved deeper into market dynamics. Despite the product's proven success in other parts of the world, they realized their IoT solution was ahead of its time for the Indian market. The infrastructure and consumer readiness in 2013 simply weren't advanced enough to support such technology.

This setback, while disappointing, taught them valuable lessons about product-market fit and timing. Rather than being discouraged, they emerged with renewed determination, actively seeking out problems ripe for disruption where their expertise could make a significant impact.

During their exploration of new opportunities, the pair found themselves increasingly drawn to the agricultural sector. Having both grown up in farming communities, they had witnessed first-hand the challenges faced by farmers in India, such as a lack of access to information, quality inputs, and modern technology. These issues resonated deeply with them, and they realized that their skills and passion could be put to use in addressing these problems. Gradually, a new focus began to take shape, and they felt they had found their true calling.

Eager to validate their new-found direction, Sachin and Sateesh spent countless hours in 2014 researching, engaging with farmers, and analyzing market trends. Their goal was to identify areas where technology could have the most significant impact and to develop a solution that would genuinely benefit the agricultural community. Through their comprehensive research and analysis, they arrived at a pivotal insight: the agri-input market in India was a vast and untapped landscape, with an estimated value of approximately $50 billion. Moreover, they identified significant gaps in the sector, such as limited access to quality inputs, a lack of transparent information, and inefficient distribution channels. This realization brought into sharp focus the immense potential for a well-crafted and strategically marketed product/service to penetrate the market and reach a substantial number of farmers. They understood that even capturing a modest 2 per cent of this market would translate to serving millions of farmers across the nation. The sheer scale and scope of the opportunity before them were both exhilarating and humbling.

Their research also yielded another critical insight: designing a product or service for farmers presented a significant challenge in terms of affordability. Most farmers could not pay for services, which meant that a way had to be found to make the venture profitable without burdening the farmers financially.

Seeking inspiration, the two looked at the strategies of successful companies that had made a significant impact in their respective fields. They were particularly drawn to Google, which had built a billion-dollar business by offering its products and services for free, yet still managed to be profitable. This example reinforced their belief that while farmers might not be able to pay for every product or service, that didn't mean a business in the agricultural sector in India couldn't be successful. As entrepreneurs, they understood that it was their responsibility to figure out how to make their venture profitable without financially burdening the farmers.

Sachin also drew inspiration from Nokia, a company that had emerged from the small country of Finland to become a global leader in mobile phones. He saw how technology could revolutionize an industry and connect people. This insight fuelled his vision for the agricultural industry, where technology could help bring transparency, connect farmers with experts, and provide access to quality products. He envisioned a platform that would offer personalized information and self-assist guides to farmers. Sachin wanted AI-enabled technologies to help farmers diagnose crop diseases and provide them with instant solutions.

As technology enthusiasts with a deep understanding of the farming community, the two young men felt they were uniquely positioned to develop products that could cater to the needs of farmers. They recognized that their knowledge of the customer base and the problems farmers faced gave them a distinct advantage in designing and developing solutions that would be both technically sound and user-friendly. Sateesh's experience as an engineering manager and global programme manager, coupled with his strong technical background, would be instrumental in planning, building the right team, preparing business plans, and formulating growth strategies. Meanwhile, Sachin's time at Honeywell had instilled in him the importance of continuous evolution and improvement, a philosophy akin to the Japanese concept of Kaizen. He understood that to scale a venture successfully, it was crucial to constantly think about what's next, what changes could be made to products or services, and what adjacent verticals could be explored to reinforce the existing systems and leverage them for growth.

Moreover, as they surveyed the competitive landscape, the entrepreneurs realized that there were no companies actively developing products specifically for farmers or the agriculture industry. This presented a significant opportunity for them to be early movers in the space, leveraging their expertise and insights to create innovative solutions that could address the unmet needs of the market.

Fuelled by this realization and their passion for making a difference, they decided to take their first leap into the agri-tech world. In late 2013, they launched their first agri-related product – an IoT-based precision water control system. They were confident that their unique blend of technological expertise and understanding of the farming community would give them an edge in the market.

However, the path to success was not as smooth as they had anticipated. Despite their thorough research and careful planning, they found that the market was not yet ready for their offering. Their product struggled to gain traction and generate sales, forcing them to confront the harsh realities of innovation in a traditional industry.

Undaunted by this setback, Sachin and Sateesh viewed the failure as an opportunity to refine their approach. They continued to explore other potential avenues for impact in the agricultural sector, brainstorming a wide range of ideas. From organic farming and contract farming to farm advisory services and the export of agricultural produce, they evaluated each concept meticulously. Their goal was to identify gaps in the agricultural sector with significant potential and large-scale opportunities that not only aligned with their strengths but also offered prospects for scalability and sustainability.

Through a process of elimination and careful consideration, they eventually narrowed their focus to the sale of seed inputs to farmers. This concept resonated with them on multiple levels – not only did it address a critical need in the market, but it also had the potential to be highly scalable and data-driven.

Their initial intuition about the potential of this idea was soon validated by external feedback. A mentor, a professor at Gandhi Krishi Vigyan Kendra (GKVK), part of the University of Agricultural Sciences, Bengaluru, conducted an informal survey among a group of farmers in his programme, asking if they would be willing to buy seeds online. The overwhelmingly positive response confirmed that there was indeed a strong appetite for such a service within the farming community.

Encouraged by this feedback, Sachin and Sateesh delved deeper into refining their concept. They realized that by combining traditional business practices with cutting-edge information technology, they could create a powerful synergy in the seed inputs space. Their vision was to leverage technology not just to streamline the sales process but also to collect valuable data on farmer preferences and behaviours. This approach, they believed, would allow them to build a highly scalable and efficient business model that could deliver significant value to both farmers and input providers.

As they further developed their idea, Sachin and Sateesh recognized an additional layer of potential in their business model. The data collected through their platform could become a powerful asset in its own right. By analyzing this data, they could gain predictive capabilities and insights that would not only inform their own strategic decision-making but could also drive innovation across the broader agricultural sector.

In developing their strategy, they also considered the existing landscape of agri-tech companies. They observed that most of the players in the field were focused on the output side of the agricultural equation, developing solutions to help farmers sell their produce more effectively. However, Sachin and Sateesh saw an opportunity to differentiate themselves by focusing on the input side, particularly on the provision of high-quality seeds.

This focus on inputs was also driven by a fundamental belief: if farmers had access to the right inputs, information, and the knowledge to use them effectively, they could significantly improve the quality and yield of their crops. This improvement, they reasoned, would have a cascading effect, leading to better outcomes not just for farmers but also for consumers and the industry as a whole.

With their vision taking shape, the entrepreneurs became increasingly convinced that they had identified a unique opportunity to make a real, tangible difference in the lives of farmers. The potential of their seed input business idea was clear – it addressed a critical need, had scalability, and could leverage data to drive innovation in the agricultural sector.

However, as the excitement of their discovery settled, the reality of pursuing this dream began to sink in, especially for Sachin. He found himself at a crossroads, facing one of the most challenging decisions of his life. As the sole breadwinner of the family, the prospect of leaving his stable job at Honeywell to start a business filled him with both excitement and trepidation. The stakes were incredibly high.

Sitting at his desk one evening, long after his colleagues had left, Sachin stared at a blank sheet of paper. On it, he began to list the potential consequences of his decision. He knew that the initial months of the venture would be gruelling, with no guarantee of a salary. This would inevitably impact his ability to support his family and maintain their lifestyle in Bengaluru, one of India's most expensive cities.

The weight of responsibility bore down on him. He thought of his children's education, his family's dreams, and the comfortable life they had built. Was he ready to risk it all for an idea, no matter how promising?

Seeking guidance, Sachin reached out to his cousin, a successful entrepreneur. 'Why are you worrying?' his cousin reassured him. 'You're only 37. If you fail, you'll have to restart your professional life after 40. That's it, nothing else. So, go for it!' While offering comfort, these words also underscored the gravity of the situation.

Sachin embarked on a thorough process of scenario-planning and risk assessment. He grappled with difficult questions: What if the business failed? How would he and his family cope with the fallout? Could he return to a traditional job if things didn't work out?

As he pondered these questions, Sachin began to see silver linings. He realized that even if his new venture were to fail, the lessons learned and skills acquired from the entrepreneurial journey would be invaluable. He took comfort in knowing that his father owned a house in his hometown of Kota, which would serve as a safety net if they needed to reduce their cost of living.

Moreover, Sachin's foresight in avoiding significant debts or financial obligations such as car or home loans during his time at Honeywell now proved to be a blessing. Unlike many of his peers who had accumulated liabilities as their incomes grew, Sachin had prioritized financial flexibility. This approach, he realized, had inadvertently prepared him for the leap into entrepreneurship.

As days passed, Sachin's resolve strengthened. The potential to create something truly transformative – a platform that could leverage technology to reshape the agricultural landscape – began to outweigh his fears. He envisioned farmers across India benefitting from their innovation, and this vision fuelled his courage.

Finally, after countless sleepless nights and deep conversations with his wife and family, Sachin made his decision to start an agri-tech company along with Sateesh. It was a moment filled with a mix of emotions – fear, excitement, and an overwhelming sense of purpose. With this leap of faith, Sachin and Sateesh officially founded BigHaat.

They knew the road ahead would be challenging, but their shared vision and commitment to leveraging technology to transform the agricultural sector gave them the courage to embark on this life-changing journey.

The genesis of BigHaat was rooted in meticulous research and a deep understanding of the agricultural ecosystem. The founders spent months in the field, engaging directly with farmers and input manufacturers. These conversations revealed a stark reality: a fragmented market plagued by interconnected inefficiencies that created a vicious cycle of challenges.

Manufacturers struggled with demand forecasting and lacked effective channels to reach farmers, often leading to overproduction or shortages of crucial inputs. They relied on multilayered distribution networks that inflated prices and reduced profit margins, making it difficult to educate farmers about new, potentially beneficial products. Farmers, on the other hand, grappled with limited access to product information and opaque pricing structures, leaving them unable to make informed decisions about which inputs would best suit their needs. This web of problems created a lose-lose situation, impacting both manufacturers' bottom lines and farmers' productivity and livelihoods.

Though recognizing that addressing these interconnected issues could revolutionize the agricultural sector, Sachin and Sateesh understood that trying to solve everything at once could lead to overextension and failure. Drawing inspiration from successful companies like Amazon (books) and Starbucks (coffee), which had started in specific niches before expanding, they decided to focus on a particular segment of the agricultural input market.

After careful consideration and market research, BigHaat chose to concentrate on horticulture seeds. This decision was driven by several strategic factors. Horticulture crops are cultivated year-round, promising a steady demand and consistent revenue stream. The sector was growing rapidly in India, with more farmers transitioning to fruit, flower, and vegetable cultivation due to the higher profit margins these crops fetched. Additionally, horticulture seeds required more technical knowledge, allowing the company to differentiate itself by providing valuable information and guidance to farmers.

By focusing on horticulture seeds, the founders believed they could make a meaningful impact while building a scalable business model. This niche would allow them to deeply understand their customers' needs, establish strong relationships with both farmers and manufacturers, and create a solid foundation for future expansion into other agricultural inputs.

However, the shadow of their previous failed tech venture loomed large, having taught them a valuable lesson: market validation should precede substantial investment. This experience shaped their approach to BigHaat. Instead of investing heavily in a complex technology platform upfront, they adopted a lean startup methodology. Their mantra became 'sell first, build later'. This pragmatic approach would allow them to focus on their core business: connecting farmers with quality horticulture seeds. By prioritizing sales and customer relationships over technological sophistication, they aimed to generate early revenues and build brand awareness without overextending their resources. The plan was to establish a solid proof of concept before investing in advanced technology. As BigHaat gained traction and validated its market fit, they would then strategically develop a more comprehensive platform, gradually refining operations and enhancing customer experience to scale their impact across the agricultural sector.

With limited personal funds, Sachin and Sateesh bootstrapped their startup, embracing frugality and efficiency at every turn. They set up operations in a small office and, alongside this, developed a basic SaaS-based e-commerce platform to support their operations. They leveraged their existing networks to minimize overhead costs. The early days were marked by relentless hustle and innovation, with the founders wearing multiple hats – from sales representatives to customer service agents. They criss-crossed rural India, personally visiting farms, attending agricultural fairs, and conducting workshops to connect with their target audience.

One of the critical challenges BigHaat faced initially was establishing effective communication with farmers. Recognizing that the cost of phone calls was a barrier for many rural customers, they implemented an ingenious missed-call system. Farmers could give a missed call to a designated number, and a team would promptly call back to take orders, note delivery details, arrange for delivery, and provide advice. This innovative approach not only overcame the cost barrier but also aligned with the farmers' current communication habits, quickly gaining traction among rural communities.

Building on this initial success, the team expanded their outreach strategy. They devised an omni-channel approach to reach and educate more farmers. Leveraging digital platforms, they utilized targeted ads on social media and search engines. However, unlike traditional advertising, their focus was not on immediate sales but on sharing valuable agricultural information and educating farmers about BigHaat's solutions. This strategy positioned the company as a trusted resource and partner for the farming community rather than just another input supplier. Complementing their digital efforts, the team conducted extensive field visits, engaging directly with farmers, understanding their needs, collecting orders in person, and building lasting relationships. It also provided them with an opportunity to collaborate with manufacturers for on-ground awareness campaigns, further solidifying BigHaat's presence in the agricultural ecosystem.

The founders established a small but dedicated farmer engagement team. This team handled customer inquiries, addressed doubts, and provided valuable agronomic advice. The personal touch fostered trust and loyalty among their growing customer base, laying the foundation for long-term relationships.

The logistics of delivering products to remote rural areas posed another significant challenge. The startup cleverly partnered with India Post, leveraging its extensive network to ensure reliable delivery of seeds to farmers across the country. This collaboration allowed them to reach even the most remote villages without investing in costly logistics infrastructure.

As deliveries increased through India Post, BigHaat encountered unexpected resistance from local middlemen. These intermediaries, who had long controlled the flow of agricultural inputs and enjoyed significant market power, viewed the new entrant as a threat to their dominance. They began to actively oppose BigHaat's deliveries, going as far as to physically block the company's delivery personnel and prevent farmers from receiving their orders.

This opposition was so severe that the startup was unable to fulfil any deliveries in February 2015, just a few months after launching its delivery operations in late 2014. The founders were forced to confront the harsh reality of the repercussions of disrupting entrenched power structures in the agricultural sector. They had to decide whether to abandon their mission in the face of opposition or find a way to overcome these challenges and press forward.

True to their entrepreneurial spirit, Sachin and Sateesh chose to persevere. They understood that the path to success would not be smooth and that they would need to adapt their approach to navigate the complex dynamics of the agricultural ecosystem. Rather than engaging in direct confrontation with the middlemen, they decided to adopt a more collaborative strategy.

They reached out to influential individuals within the existing system, such as progressive farmers and industry veterans, and sought to engage them as partners in their mission. By working with these stakeholders and leveraging their networks and influence, BigHaat could gradually build trust and acceptance within the farming communities they served.

This experience taught the founders a valuable lesson: collaboration and communication, rather than competition, are the keys to driving change in the agricultural sector. They realized that by working within the existing system and building bridges with key stakeholders, they could slowly but surely transform the landscape and create a more equitable and efficient ecosystem for all.

With this new-found understanding, BigHaat resumed operations in March 2015, adopting a more measured approach to growth. Their commitment to providing high-quality products, fair prices, and valuable agricultural knowledge slowly but surely gained traction. Word of their positive impact spread, attracting more farmers to the platform.

This growing success caught the attention of major industry players. In a significant milestone, US Agriseeds, a leading agricultural company, approached the startup, expressing interest in partnering with them. This was a major validation of BigHaat's business model and potential, especially considering that just a few months earlier, an experienced agriculture expert had predicted that the startup would not survive beyond a quarter.

The partnership with US Agriseeds marked the beginning of a new chapter for BigHaat, providing them with access to a wider range of high-quality products, industry expertise, and distribution networks. This collaboration helped accelerate the startup's growth and expand its impact, reaching more farmers across India.

Within the first year of operations, BigHaat's website attracted over 35,000 farmer visitors seeking information and solutions, a testament to the platform's growing influence in the Indian agricultural landscape. As orders increased and their business grew, Sachin and Sateesh faced the challenge of upgrading their existing basic SaaS-based e-commerce platform to a more robust and scalable website.

They carefully evaluated their options, considering the pros and cons of the alternatives: building an entirely new platform from scratch, customizing an open-source e-commerce framework, or migrating to a more comprehensive third-party solution. Building from scratch or heavily customizing an open-source platform could provide more tailored functionality but would require significant time and resources for development and ongoing maintenance.

Sachin and Sateesh ultimately decided to go with a ready-made e-commerce solution to keep their monthly expenses low. This strategic decision allowed them to focus their efforts and resources on marketing, customer acquisition, and order fulfilment rather than getting bogged down in complex website development and maintenance. The chosen platform provided a solid foundation for BigHaat's online presence, giving them the flexibility to customize and scale as the business grew.

This decision highlighted Sachin and Sateesh's pragmatic approach to building their startup. While many entrepreneurs might be tempted to invest heavily in creating a cutting-edge, feature-rich website from the outset, BigHaat's founders understood the importance of prioritizing their limited resources. By opting for a cost-effective, ready-to-use solution like Shopify, they could direct their energies towards optimizing their business model, acquiring customers, and generating revenue. This lean, iterative approach would serve them well as they navigated the challenges and opportunities of the rapidly evolving agri-tech landscape.

One of the key decisions the founders made early on was to position BigHaat as a farmer-centric platform. While their initial focus was on horticulture seeds, they recognized that their platform had the potential to serve multiple stakeholders in the agricultural value chain, including manufacturers and retailers. However, they understood that trying to cater to all these stakeholders simultaneously with limited resources could dilute their efforts and impact.

After careful deliberation, Sachin and Sateesh concluded that farmers should be their primary customers and the focal point of all their business decisions. They believed that by prioritizing the needs and challenges of farmers, they could create a digital agriculture ecosystem that truly made a difference in their lives. Manufacturers and retailers, while important partners, would be considered enablers in delivering value to farmers rather than the primary beneficiaries of BigHaat's services.

This farmer-centric approach would go on to define the organization's culture and strategy, influencing every aspect of its operations, from product development to customer support. By aligning their efforts towards empowering farmers and improving their livelihoods, the founders could build a digital agriculture ecosystem that not only addressed the immediate needs of their customers but also contributed to the larger goal of transforming the agricultural sector in India.

As BigHaat continued to make rapid strides, it validated the founders' confidence in their venture. In the early days, industry experts raised their eyebrows, questioning the viability of an online platform for agricultural inputs. 'Farmers won't adopt digital solutions,' they said. 'It's a pipe dream,' others scoffed. But where critics saw insurmountable obstacles, Sachin and Sateesh saw golden opportunities. They recognized a crucial fact that others had overlooked: farmers were already familiar with e-commerce platforms like Flipkart and Amazon for their household needs. Why not extend this digital comfort to their professional lives? The founders believed that with a user-friendly interface, reliable delivery, and high-quality products, they could bridge the gap between traditional farming and modern technology.

Critics argued that only a minuscule 2–3 per cent of farmers would ever use such a service. Sachin's response was nothing short of revolutionary: 'Do you realize the power of that 2 per cent?' he'd ask. In a market as vast as India's $50 billion agricultural input sector, even a small percentage translated into a billion-dollar opportunity. It was this ability to see the forest for the trees that set these entrepreneurs apart.

The naysayers didn't stop there. They predicted that BigHaat's impact would be limited to a handful of tech-savvy farmers, never gaining widespread adoption. But the founders understood the domino effect of positive change. They believed that early adopters would experience improved productivity and quality of life, becoming walking testimonials for the company. And they were right. As word spread from farmer to farmer, family to family, village to village, the user base grew exponentially.

Another ace up their sleeve was Sachin's background in telecom. While others saw India's digital infrastructure at that time as a limitation, Sachin saw it as a wave of opportunity about to crest. He anticipated the rapid expansion of 4G networks, making mobile internet more accessible and affordable for farmers. This foresight allowed BigHaat to position itself at the intersection of agriculture and technology, ready to ride the digital wave when it arrived. The founders didn't just address the market as it was; they envisioned what it could become.

This visionary approach paid off handsomely. In December 2016, the startup secured its first round of funding, a testament to the strength of the founders' business model and the trust they had built. The following year saw the company reach new heights, with sales and traction peaking in what would be remembered as a landmark year for the company.

However, the path of entrepreneurship is rarely smooth. In 2018 and 2019, BigHaat faced significant setbacks when two anticipated investment opportunities fell through, leaving them with limited financial resources to fuel further growth. But true to their nature, Sachin and Sateesh saw these challenges not as roadblocks but as opportunities to refine their strategy and prove their resilience.

September 2019 marked a particularly challenging period for BigHaat, as the company found itself at its lowest point since its inception. With limited funds to pay salaries and settle vendor dues, the founders found themselves in uncharted territory, unsure of the path forward.

Friends and colleagues who had witnessed their unwavering commitment and the potential of their platform stepped forward to support them when they needed it most. These individuals believed in Sachin and Sateesh's integrity, resilience, and ability to overcome the challenges they faced. They provided financial support, investing their own money and rallying their networks to help BigHaat weather the storm.

One key figure in this journey was Kiran Vannum, who joined Sachin and Sateesh as a co-founder, sharing their vision for BigHaat. Kiran's role was crucial, focusing on expanding the company's presence in the states of Andhra Pradesh and Telangana. His dedication and performance significantly contributed to the company's growth and resilience. Kiran's belief in the BigHaat story was so strong that he mobilized his friends and relatives to invest in the company during its time of need, providing it vital support when it was most critical.

The three co-founders stood together during tough times, their shared belief in BigHaat's mission and potential keeping them united. The period of difficulty they experienced and emerged from underscored the importance of building a strong founding team with genuine relationships based on shared values and a common vision. Together, they weathered the challenges, each playing a vital role in shaping the company's journey from its inception.

As the startup entered 2020, the company's fortunes began to turn around. Beyond Next Ventures, a Japanese venture capital firm, recognized the potential of BigHaat and invested in the company, providing a much-needed influx of capital. This investment not only helped stabilize its financial position but also allowed for future planning and investment in strategic initiatives. This vote of confidence from an international investor marked a turning point for the company.

Simultaneously, the company was expanding its horizons. After three years of focusing solely on seeds, it made a strategic pivot to diversify its product portfolio. The team gradually introduced crop protection solutions, nutrients, and farm implements, transforming itself into a one-stop shop for farmers' agricultural needs.

The year 2020 also heralded a significant shift in India's telecom and technology landscape. The widespread adoption of 4G networks and increasing smartphone penetration in rural areas created fertile ground for agri-tech solutions to flourish. The founders, with their characteristic foresight, recognized that the time was ripe to double down on technology and build a robust digital infrastructure to support growth.

Interestingly, this realization coincided with the onset of the COVID-19 pandemic, presenting both challenges and opportunities. While many companies focused on cutting costs and conserving cash during the crisis, Sachin and Sateesh took a contrarian approach. They believed that investing in technology during this period could give BigHaat a competitive edge and position the company for long-term success.

However, BigHaat's financial reserves were limited, making it difficult for them to allocate substantial resources towards technology development. Sachin and Kiran explored innovative partnerships and collaborations to bridge this gap. They identified a technology partner who believed in their vision and was ready to work with limited resources in co-developing cutting-edge technology solutions for the agricultural sector.

Prior to this, BigHaat had a basic app with a web interface at the back end. However, in October 2020, with the help of its technology partners, it launched its first true native app, marking a significant milestone in the company's digital transformation journey. The app provided a seamless and intuitive user experience for farmers, empowering them to access a wide range of products and services at their fingertips.

To ensure that the app remained user-centric and responsive to the evolving needs of farmers and other stakeholders, the founders implemented a robust feedback mechanism. The team actively sought inputs from farmers, analyzing their behaviours, preferences, and pain points to continuously improve the app's features and functionality. This iterative approach allowed the company to release new updates and enhancements every fortnight, keeping the platform fresh, relevant, and aligned with the needs of its users.

As BigHaat's digital presence grew, so did its influence in the agri-inputs supply chain. Manufacturers began to take notice of the platform's potential to disrupt traditional distribution channels. Initially, when BigHaat opened an account with US Agriseeds, only a handful of manufacturers came on board. The team found themselves putting in long hours, tirelessly pitching the benefits of e-commerce to sceptical suppliers. Many manufacturers were hesitant to break away from the tried-and-true methods of the past, unwilling to risk their products on what they perceived as an unproven digital upstart.

But the team refused to be discouraged. If the big players wouldn't come to them, they would build from the ground up. They turned their focus to forging strong relationships with local distributors and retailers, leveraging their on-the-ground networks and deep market expertise. Slowly but surely, product offerings began to expand as more and more local partners bought into their vision of a digital future for agriculture.

As sales figures climbed and consistency in order deliveries shone through, manufacturers began to take a second look. The writing was on the wall: e-commerce was here to stay, and BigHaat was positioning itself as a serious contender in the agri-tech space. Suppliers started to recognize the vast potential of partnering with the platform, envisioning access to untapped markets and new customer segments.

Gradually, a shift began to take hold. The same manufacturers who had once given BigHaat the cold shoulder were now proactively reaching out, eager to establish partnerships and supply their products directly through the platform. It was a remarkable turnaround, a testament to the tireless efforts of the founders and their team in proving the value of their model.

With BigHaat's growing influence and the shift in manufacturer sentiment, the co-founders knew it was time to take their platform to the next level. They turned to Kiran Vannum to spearhead a ground-breaking new initiative: market linkages. The aim was to create a full-stack solution that would revolutionize the way farmers interacted with the agricultural value chain.

The concepts of market linkages and full-stack solutions were relatively new in the agri-tech space at the time, and most farmers were unfamiliar with their potential benefits. In simple terms, a full-stack solution refers to a comprehensive platform that covers all aspects of the agricultural value chain, from the provision of quality inputs and advisory services to connecting farmers with buyers and facilitating seamless transactions. By enabling these linkages and closing the loop, BigHaat aimed to create a more efficient, transparent, and profitable ecosystem for all stakeholders involved.

Before launching the market linkage programme, the co-founders engaged in deep discussions to identify the key value additions they could offer to farmers. Staying true to their farmer-centric approach, they realized that helping farmers follow best practices to grow quality crops and obtain access to export markets could be a game changer. The company was already providing high-quality inputs and advisory services to help farmers improve their yield and output quality. By taking it a step further and assisting farmers in meeting the stringent requirements of export markets, they could unlock new opportunities for growth.

The founders knew that success in this venture would hinge on two critical factors: trust and transparency.

Farmers needed to have unwavering confidence in the platform, and buyers needed to know they could rely on the quality and origin of the produce they were purchasing.

To address this, the team introduced a cutting-edge traceability system that would become the cornerstone of their market linkage programme. This innovative solution allowed buyers to access detailed information about the entire cultivation process, from seed selection to harvest and post-harvest handling. By providing this level of transparency, BigHaat could foster long-term relationships, built on trust, between farmers and buyers.

The company took its commitment to transparency and sustainability even further with the launch of the farmer-centric Sustainability Development Programme (SDP). Working hand in hand with farmers, the SDP aimed to promote the adoption of sustainable farming practices, maintain rigorous quality standards, and assist farmers in meeting the necessary certifications for export. The company recognized that even if the farmers' output didn't always meet export criteria, the focus on quality and sustainability would still result in significantly higher-grade produce for the domestic market.

By leveraging cutting-edge technology, BigHaat could provide buyers with real-time data on the farming practices employed, the inputs used, and the quality checks conducted at various stages of the crop cycle. This level of granular detail helped build a strong level of trust among the farming community and foster long-term relationships between farmers and buyers.

Behind the scenes, BigHaat was building a sophisticated agri-data stack that collected, analyzed, and leveraged vast amounts of data to drive efficiency, optimize operations, and create value for all stakeholders. The platform's data capabilities enabled personalized recommendations for farmers, predictive analytics for demand forecasting, and intelligent supply chain management.

In addition to the market linkage programme, the company continued to expand its digital infrastructure and offerings to empower farmers and enhance their livelihoods. BigHaat launched Kisan Mudra, a fintech linkage product under its platform. This solution connected farmers with fintech partners to meet their financial needs, providing easy access to credit and financial services. This innovation eliminated traditional barriers and reduced paperwork associated with agricultural loans, making financial resources more accessible to farmers.

Recognizing the importance of knowledge-sharing and community building, BigHaat introduced Kisan Vedhika, a digital agri-community platform that connected farmers, experts, and other stakeholders, fostering collaboration, learning, and collective problem-solving. Farmers could interact with each other, share best practices, and seek advice from experienced professionals, creating a vibrant and supportive ecosystem.

Another innovative offering was Crop Doctor, an app under the BigHaat digital platform that leveraged advanced artificial intelligence and machine-learning algorithms to provide intelligent solutions to farmers. One of its key features was its ability to diagnose crop diseases and provide actionable recommendations based on a simple photograph uploaded by the farmer. This technology not only helped farmers quickly identify and address crop health issues but also enabled BigHaat to generate real-time alerts and advisories for neighbouring farms, helping prevent the spread of diseases and improve crop yields and quality.

As the digital infrastructure and data capabilities grew, so did the impact on the farming community. The platform's ability to deliver personalized, data-driven solutions at scale was transforming the way farmers approached their work, empowering them to make informed decisions, reduce risks, and maximize their potential.

One of the key factors contributing to BigHaat's success was the founders' unwavering commitment to building a long-term, sustainable business. They were not in it for short-term gains; instead, they focused on creating a platform that could stand the test of time and deliver lasting value to all stakeholders. Their complementary skills created a well-rounded and resilient organization. Sateesh handled strategy, stakeholder management, and technology development while also crafting the future roadmap for the company. Kiran focused on market linkages, working closely with farmers to grow quality crops and collaborating with national and international buyers. Sachin concentrated on input linkages, engaging and serving farmers at a pan-India level while also collaborating with and bringing together all stakeholders to create a meaningful impact on the lives of farmers.

The three partners believed that the quality of their work should speak louder than any advertisement. They focused on delivering exceptional products, services, and experiences to their customers, trusting that the positive impact they created would organically attract more users and partners to the platform. This approach was evident in their marketing strategy, with BigHaat ads being a rare sight in the media landscape.

Another crucial factor that contributed to BigHaat's resilience and growth was the strong bond of trust and mutual support among the three founders. Even during the challenging times of 2019, when the company faced a financial downturn, they stood together, focusing on finding solutions and ways to move forward. Their belief in each other and their shared vision helped them weather the storm and emerge stronger.

Sateesh, in particular, stood out as a people-centric leader, always prioritizing the well-being and growth of the team. Even when the company faced financial constraints, he ensured that employees were paid on time, even if it meant delaying payments to vendors or to the founders themselves. During the COVID-19 pandemic, Sateesh's leadership and compassion shone through as he made sure that no team member lost their job, recognizing the importance of supporting his people during challenging times.

BigHaat's business model centred on providing a comprehensive digital platform for farmers. The company generated revenue through multiple streams:
1. Sales of agricultural inputs (seeds, fertilizers, crop protection products)
2. Commission from market linkages (connecting farmers with buyers)
3. Data-driven services and insights for agribusinesses

The focus on technology and data-driven solutions allowed for efficient operations, keeping costs low while scaling rapidly. This approach enabled BigHaat to achieve profitability within five years of operations, with revenues growing at over 100 per cent year on year for three consecutive years.

By 2023, the platform had connected over 10 million farmers across India. A majority of these farmers were small and marginal landholders from rural areas, with a growing number of tech-savvy young farmers joining the network. The user base spanned 15 states, with a strong presence in major agricultural regions.

The founders' vision for BigHaat was ambitious yet grounded. They aimed to build the largest food-safe company in India by working closely with farmers and leveraging technology to transform the agricultural landscape. While dreaming big, they were mindful of growing at a sustainable pace.

Most importantly, Sachin, Sateesh, and Kiran were building BigHaat not just for themselves but for the future. They wanted to create a company that could thrive and continue to create value, a global digital agri ecosystem that could stand the test of time.

# 2

# The Intuitive Innovator: Prateep Basu, SatSure

*A childhood fascination with the cosmos, sparked by the tragic Columbia disaster and Kalpana Chawla's heroic story, set Prateep Basu on a path few would dare to travel. From the hallowed halls of ISRO to the frontier of private space innovation, Prateep seemed destined for a career among the stars. But his analytical mind couldn't ignore the disconnect between India's agricultural sector and the revolutionary potential of satellite technology. Rejecting the comfortable certainty of an established career, Prateep ventured into the uncharted territory of space science meeting farming – a domain sceptics deemed too complex and investors considered too risky. Through a bootstrapped beginning that tested his limits, he built SatSure from a mere concept to a pioneering force that now aids banks in making better lending decisions for farmers. His story isn't merely about technological innovation but about reimagining how satellite imagery can democratize financial access for millions of India's farmers, proving that sometimes the most impactful application of cutting-edge technology is found not in exploring new worlds but in transforming our own.*

The date was 1 february 2003, and 24/7 news channels were the latest craze in the small towns of India. Never had the country felt so connected before, especially with global events. Thanks to this new phenomenon, the nation learned about the tragic accident involving the space shuttle that claimed the life of Kalpana Chawla, one of six astronauts returning to Earth from an international space station.

Prateep Basu was only thirteen years old at the time, and the story of Kalpana Chawla's journey from a small town in Haryana to NASA, where she became an astronaut, left a lasting impression on him. As a child, he was always captivated by the wonders of the universe and the limitless possibilities of science fiction. However, there weren't many people around him in Ranchi, his hometown, who shared his enthusiasm for space and astronomy. Finally, he had someone to look up to, someone who would eventually become an inspiration in his career path a few years later.

His early fascination with outer space was sparked by regular visits to the British Library in Ranchi with his elder sister. The works of renowned authors like Isaac Asimov, Arthur C. Clarke, and H.G. Wells ignited a passion in him for the subject that would shape the course of his life. As a young boy, he found himself enthralled by the terrifying saucer-like alien ships in the 1995 movie *Independence Day*. Watching it on cable television made him curious about the prospect of extraterrestrial life, about humanity's exploration of other planets, and the potential consequences of human contact with alien civilizations.

The young enthusiast would read his elder sister's physics books to learn about planetary formation and astronomy, as there weren't any other resources at his disposal to quench his curiosity about the unknowns of deep space. However, the corny sci-fi movies of the 1980s and 1990s, like the *Terminator* series, *Armageddon*, *Alien*, and *Predator*, also played a significant role in fueling his imagination and his interest in space and technology.

In 1998, his parents got a personal computer at home for his sister, which accelerated his quest for answers to the questions that plagued his mind. Yahoo and Google became his favourite websites for the next few years! Prateep's parents, while encouraging him to pursue academic excellence, never pressured him to be at the top of his class or achieve perfect grades. He was always active in extracurricular activities like sports, the arts, and quiz contests. His parents fostered a well-rounded upbringing, allowing him to devote equal time to his studies and the exploration of his creative passions, something he continued throughout his college years.

Despite his involvement in extracurricular activities, sports, and quiz contests, Prateep excelled academically too. He topped his school in the Class 10 board exams and was expected to make a career decision right away, as the subjects one studied over the next two years defined at least two things for science students back in the early 2000s – whether they would choose medicine or engineering!

While his parents harboured dreams of him becoming a doctor, his own interests lay elsewhere. Prateep's fascination with space had fostered a deep love for physics and its mathematical interpretations. This passion drove him to seek out opportunities to solve problems and build things, leading him to consider a career in engineering over medicine.

Determined to pursue his interests, he began preparing for the Joint Entrance Examination (JEE), the gateway to securing a spot in one of the prestigious IITs. He threw himself into his studies, spending long hours mastering complex concepts and honing his problem-solving skills. However, despite his best efforts, his rank of 3,000+ in the exam fell short of the cut-off for admission to an IIT.

As he watched his friends celebrate their success and prepare to embark on their engineering journeys at these renowned institutions, the young aspirant couldn't help but feel a sense of disappointment and uncertainty about his own future. But thanks to his active participation in sports, he had learned the valuable lesson that one must be prepared for both triumph and defeat in life.

It was during this period of contemplation and self-doubt that a serendipitous email arrived, introducing him to the Indian Institute of Space Science and Technology (IIST) in Trivandrum, a newly established government educational institute backed by the Indian Space Research Organization (ISRO). The name alone was enough to pique his interest, and without hesitation he applied, knowing little about the college or its offerings.

To his surprise, an invitation to a counselling/seat-selection session in Bengaluru followed. There, he and his parents had the privilege of meeting senior ISRO scientists, some of whom were esteemed Padma Shri and Padma Bhushan awardees. It was also there that he discovered the renowned scientist and former President of India, Dr A.P.J. Abdul Kalam, a figure he had long admired for his role in India's space and missile programmes, was the chancellor of the college.

The opportunity to study at an institution led by such high-profile professionals and to pursue a degree in aerospace engineering felt like a dream come true. From a perceived failure, he got the chance to learn from and work alongside some of the brightest minds in a field that had fascinated him since childhood.

IIST, the first university in Asia dedicated solely to the study and research of space technology, was established in 2007 to meet the needs of scientists and engineers in the Indian space programme. When Prateep joined, the university was still in its nascent stages, operating from a temporary building within the Vikram Sarabhai Space Centre (VSSC), Trivandrum. Despite the challenges faced by the young institution, it provided Prateep with a nurturing environment where he could develop a wide array of hard and soft skills.

The young man had the rare opportunity to witness the birth and growth of a university. He saw from the front row what it took to execute the bold vision of the founding director of IIST (and current chancellor), Dr B.N. Suresh, who actively involved the students in growing the organization. He also witnessed first-hand the benefits of encouraging people to think and act like owners, as he and his classmates took on increased responsibilities and generated innovative ideas for the institute's development. In many ways, his college experience mirrored that of a startup, providing him with valuable entrepreneurial insights. Today, IIST is recognized as one of the fifteen institutes of national importance and has produced over 2,000 engineers for ISRO and for the outside world.

Prateep's time at IIST proved to be a transformative learning experience, equipping him with technical knowledge and hands-on exposure to various laboratories. He had the opportunity to work on real projects in collaboration with units of ISRO, further enhancing his skills and expertise. However, as graduation approached, he and his classmates found themselves grappling with concerns about their future prospects. The university had been established with the primary objective of supplying ISRO with talented graduates, and the realization that finding employment elsewhere might be challenging if they failed to secure a position at ISRO weighed heavily on their minds. Fortunately, Prateep's grades were sufficient to ensure his placement at ISRO.

During the placement process, the graduates were asked to select a job at one of the ISRO centres. To the surprise of many, Prateep opted for Sriharikota, a remote island that was not a popular choice among the other students and was even perceived by some as a form of 'punishment posting'. Given his impressive research profile, his decision raised eyebrows, even among his friends.

The young man's primary motivation for selecting Sriharikota was his desire to gain as much experience as possible in the shortest time frame. He believed that the faster he accumulated experience, the more rapidly he could advance his career. Additionally, having already completed a four-month project at the ISRO unit of the Liquid Propulsion Systems Centre (LPSC) in Trivandrum, he was well acquainted with the work environment there and at VSSC. Prateep had no intention of spending his days confined to a desk job doing research like running simulations that took weeks to yield results. At Sriharikota, he would be in close proximity to the launch pad, enabling him to witness the direct impact of his efforts and interact with the project teams for both launch vehicles and satellites, as everything culminated at the launch site.

With this goal in mind, the aspiring scientist embarked on his career at ISRO. He was 22 at the time. There he contributed to the ongoing development of the GSLV MK-III (Geosynchronous Satellite Launch Vehicle Mark 3) project and participated in 11 successful PSLV (Polar Satellite Launch Vehicle) launches. However, after only a couple of years, he began to sense that his growth at ISRO had plateaued, and his learning rate had diminished to the point where he felt he was no longer acquiring new knowledge.

The young professional reached out to his friends from school who were working in the private sector and realized that such stagnation early in his career could be detrimental to his future. With limited options available to him, Prateep contemplated a return to academia to pursue a master's degree or an MBA. However, he recognized that his experience in the space industry might not be directly applicable in other fields, and he risked being treated as a fresh graduate.

Faced with these challenges, the ambitious young man made the bold decision to resign from his position at ISRO. However, he was bound by a five-year bond and breaking it would require him to pay a substantial sum of ₹10 lakh.

Lacking the personal financial means to do so, he approached his father and persuaded him to view the bond money as an investment in his future. With his father's support, Prateep was able to leave ISRO and pursue a master's degree in space science at the International Space University (ISU) in France, where he was awarded a full scholarship.

While leaving his job at ISRO was a risky move, it ultimately proved to be a rewarding decision that allowed him to continue his learning journey, both professionally and personally. He joined ISU at a time when space technology companies were starting to boom. The university served as an incubator for space technology entrepreneurs, providing Prateep with the opportunity to experience once again a startup work culture and gain insights into interdisciplinary thinking and venture building. It was during this period that he first learned about the commercial space industry and its immense size and potential.

As he was completing his master's degree, he embarked on the challenging task of securing employment. Despite applying for over 120 job positions, only two organizations extended him invitations for interviews. Fortunately, after several rounds of interviews, both companies offered him a position. One was a well-established, large company based in Luxembourg, boasting generous salaries and benefits. However, Prateep opted for the road less travelled, choosing to join a small research and consulting firm that offered a lower salary but the option of remote work (much before it became a thing!).

It was a calculated risk, driven by his belief that this opportunity would provide him with greater flexibility, learning opportunities, and the chance to work closely with C-suite executives. He returned to India in July 2015 with a job that required him to be equipped with only a laptop, phone, and a good internet connection to succeed. While he could have literally worked from home, which was then Kolkata for him as his family had moved there, he chose to begin a new life in Bengaluru – the IT city of India, which was also the aerospace city of India.

Prateep's decision to move to Bengaluru was part logic and part bet. It was logical because his job entailed frequent travel across the world, and Bengaluru was well connected. The city also had several critical ISRO units and other national and private aerospace organizations. And to top all this, he had quite a few good friends from IIST who were also working in the city, so he had a social circle there too. But it was a bet because he believed Bengaluru could be what San Francisco was for the emerging private space industry, commonly known as 'NewSpace'.

As he had anticipated, his role at the consulting firm allowed him to interact with numerous venture capitalists, tech entrepreneurs, and CXOs, enabling him to gain invaluable knowledge and expertise. During his consulting years, Prateep had the opportunity to engage with several companies that operated satellites, performing earth imaging, and building applications utilizing satellite data. These companies, primarily based in the United States and Europe, provided services to their respective governments for several applications in defence and intelligence, disaster management, oil and gas, and the agricultural sectors.

While he had been aware of the societal applications of satellite data since his undergraduate days at IIST, Prateep had to apply himself to understand remote sensing and earth observation, as he came from an engineering background. His interactions with senior personnel from these companies provided him with insights into the commercial use cases of satellite data.

Among the various types of end users, it was the farmers to whom his attention was drawn. He had come across news about farmers getting paid meagre amounts as compensation for crop loss resulting from climatic events. While he was suggesting business models for using satellite imagery for such applications to companies globally, right in his backyard both farmers and the government seemed oblivious to the use of technology for improving farm insurance and credit. As he aptly put it, 'Farmers are the riskiest entrepreneurs. Every season, they work without knowing if they will earn money. They cannot predict market demands, rising costs, price variability, or natural events.'

Inspired by his findings and driven by a desire to make a difference, the young consultant began researching ways to leverage satellite data to assist farmers, all while continuing his regular work. He engaged in discussions with farmers, bureaucrats, agribusiness professionals, agricultural investors, and cooperatives to gain a deeper understanding of the challenges they faced, including user behaviours, environmental factors, supply chain issues, and financial impediments.

Through his research, he discovered that farmers were often sceptical of technology and found it challenging to view and interpret information. He realized that for satellite data to be truly useful to farmers, it needed to be transformed into something they could easily comprehend, accompanied by actionable recommendations. Already in the US and Canada, firms had started providing such services to farmers. However, after careful deliberation, he concluded that it might not be possible to serve Indian farmers directly using such technology. Not only were farm holdings very small in India, leading to the interpretation of satellite imagery not being consistently accurate enough to be brought directly to farmers as advice, but the paying capacity of farmers for 'soft services' was also very low.

Despite this finding, the aspiring entrepreneur recognized that satellite imagery's unparalleled ability to collect data at scale, create interventions, and monitor risks would have an impact when it came to enterprises serving farmers. This led him to consider an alternative approach: instead of focusing on providing advice directly to farmers, he could use satellite data to enhance risk perception and monitor various agricultural stakeholders, including insurance companies, policymakers, banks, traders, and agribusinesses. By doing so, he could indirectly support farmers by improving the decision-making processes of the institutions and individuals that impacted their lives.

This prompted Prateep to contemplate the establishment of a business whose core value proposition would be risk monitoring using satellite imagery. He envisioned a company that could leverage the large datasets generated by the imagery to provide insights and tools that would help agricultural stakeholders better understand and manage the risks they faced, ultimately leading to informed decisions and better outcomes for farmers.

Although he was refining his business idea through extensive research and deliberation, the young professional did not feel an immediate urgency to start a company. He was enjoying a comfortable and privileged life, earning a substantial income for someone in his mid-twenties and loving the new-found upward social mobility. However, a serendipitous encounter would soon change his perspective.

One day, while accompanying a friend to a photo shoot in Bengaluru, he found himself sharing an Uber ride home. During the journey, he couldn't help but overhear the driver's tense phone conversation. The driver was expressing his urgent need to complete a certain number of trips before midnight to acquire the necessary funds for something. Intrigued by the situation, Prateep gently inquired about the driver's predicament. The driver, initially hesitant, eventually opened up and explained that his mother had to undergo surgery and he was desperately trying to earn the shortfall of money they faced by making additional trips. Despite having collected some funds for the operation, he was still struggling to come up with the rest.

Moved by the driver's plight, the young consultant offered him financial assistance to help cover the cost of the surgery. The driver, though reluctant at first, eventually accepted the offer. A week later, to Prateep's surprise, the driver called him to return the money and actually turned up at his place! Intrigued by this gesture, he invited the driver to a nearby coffee shop for a friendly conversation.

As they sat down and began to talk, the driver's story unfolded, revealing the challenges his family had faced over the years. The driver's family had been involved in sugarcane farming in Mandya, a place 90 km from Bengaluru. But a dispute with a sugar mill owner over borrowed money had taken a devastating turn when goons destroyed his crops and gave a communal angle to the incident. In an effort to recover from this setback, the family borrowed money from a private lender and ventured into sericulture. However, their struggles were far from over. Two consecutive droughts destroyed their crops and worms, leaving the family in a dire financial situation.

Burdened by mounting debts and the relentless pressure from the moneylenders to pay up, the driver's father had tragically taken his own life, leaving the family in a precarious position. Desperate for a better future, they had moved to the city, hoping to find employment and pay off their debts. But just as they were trying to get back on their feet, the driver's mother fell ill and required surgery, adding to their already overwhelming troubles.

As Prateep listened to the driver's heartbreaking story, he was deeply moved. The driver's tale of hardship, loss, and resilience in the face of adversity was a stark reminder of the challenges faced by countless farmers and their families across the country. This turned out to be the 'tipping point' for the young professional, prompting him to take the leap into entrepreneurship, as he knew that maybe his business idea could make a difference in the lives of those who struggled in similar situations. The driver's story highlighted the genuine and pressing need for change in the way agriculture was financed and how farmers were supported in times of crisis.

The aspiring entrepreneur was aware that ideas are more powerful than individuals, and it was abundantly clear to him that there needed to be a structural change at the institutional level. He needed a team that would buy into this vision, and that's when he reached out to his friend Abhishek Raju, who was already an entrepreneur and part of a small satellite manufacturing startup. Abhishek suggested that he pitch his business concept to the entrepreneurship cell of IIM Bengaluru, as it was an up-and-coming hub for nurturing early-stage startups.

Hoping to secure the support and resources needed to get his venture off the ground, Prateep made his first pitch deck – a disaster according to him as he looks back! Despite his best efforts, his pitch fell short, and the investor team swiftly rejected his proposal. However, the idea that farmers needed better access to financial services so that they could protect themselves and their families against climate risks resonated with one of the jury members. This person approached him after the pitching session, expressing interest in investing in the business. He remarked, 'I liked your pitch. It needs a lot of thought processes to explore how to make money. I will give you five lakh rupees for a 20 per cent stake.'

Stunned and amused at the same time by such an offer, the young innovator declined it, but the fact that someone was willing to invest in his idea served as validation of its potential. Fuelled by a renewed sense of purpose, he decided to take the first step towards bringing his idea to life. He quit his comfortable job.

He had this angel investor to thank for finally taking the plunge, as the advice he received from him stuck with

Prateep: 'If you are truly interested in your idea, then you should quit your job and devote yourself to it full time. Have real skin in the game.' Prateep reached out to Abhishek and convinced him to partner with him to build on his idea and create a full-fledged company. This marked the birth of SatSure, a company whose value proposition was to assist farmers in obtaining better loans and faster insurance claims, thereby promoting financial inclusion and helping farmers break free from the debt trap of moneylenders.

Having a co-founder was great, but still the budding entrepreneur needed to assemble a team of individuals who would be able to execute his vision and add to it because he didn't have the technical skill sets himself. It wasn't so difficult to identify where to find such people, given his background in the space sector. So he reached out to a few of his friends from college who had studied and worked in the domain of satellite remote sensing. First up were Rashmit and Bharat, both good friends of Prateep's and people whom he had already confided in about the idea of SatSure.

Rashmit was finishing his MBA in Canada, but because of his background in remote sensing, he was quick to show interest in building small proofs of concept for Prateep to demonstrate the capability of their idea to people he met. Bharat was working at ISRO at the time and was unsure about how he could contribute as there was no clarity among any of them as to how commercially viable the idea of SatSure was at that time. He decided not to join and went to Germany instead to pursue his master's in satellite technology.

Abhishek meanwhile helped Prateep connect with several industry professionals, and both met a bunch of people to seek help in building a minimum viable product. It was mid-2017, and the term space-tech was unheard of by VCs in India, while agri-tech was a newly coined term that was gaining interest and momentum. So the founders initially fashioned SatSure as a novel agri-tech firm and pitched the concept to large insurance companies. However, to their disappointment, these companies were hesitant to take on the technology risk as the 'first customer'. Some of them also confused the offering with weather-risk forecasting, but Prateep thought this was probably due to his poor pitching!

Undeterred, the founders turned to the Government of India to get their business idea validated and to give them that first opportunity to demonstrate their concept. This was the time when schemes like Startup India and Make in India were making waves in the Indian startup ecosystem. Abhishek's friend and associate from another venture, Sravan, was a young college graduate trying to build a political career. He offered his assistance in securing some meetings.

After some perseverance from Sravan, Prateep secured 15 minutes of 'talk time' with a young member of Parliament from the Srikakulam district of Andhra Pradesh, Shri Ram Mohan Naidu. A Purdue University graduate, Naidu came from an eminent political family, yet he didn't come across to Prateep as a typical politician. He gave a patient hearing to Abhishek and Prateep's idea, asked a few very relevant questions about validation and impact, and then did that one thing no one else did – he gave them a chance to pilot their idea in his district Srikakulam, a paddy bowl often affected by cyclones.

Prateep and Abhishek were then in a fix – they had this golden opportunity but not the people to execute it within the time frame that made it meaningful for everyone. They needed to build a team immediately! Prateep convinced Rashmit to join as a co-founder. He returned to India, leaving behind a well-paying job offer, while also enrolling a couple of other folks from his circle of friends to start the work in Srikakulam.

The letter of intent provided by Shri Ram Mohan Naidu led to Prateep and Abhishek registering a private limited company. Next was identifying how to finance the initial months of operations, as the pilot project in Srikakulam was being done on a no-cost-no-commitment basis. Both had some savings and some friends, one of the most important among whom in the early days was Dr Roger Moser, a Swiss professor at the University of St Gallen whom Abhishek had known for many years. It was important at that time to create cash flows in the company to be able to hire at least three or four people. Abhishek, with help from Roger, activated his networks in Switzerland and sent Prateep to Zurich to close the first commercial deal for SatSure.

Funnily enough, this first commercial deal for SatSure was neither related to satellite imagery nor agriculture. It was a purely opportunistic service that Prateep had no clue about – summarization of PDF documents for an asset management company using natural language processing. The money was good, and it would help support salaries for a few people for at least nine months, so Prateep was determined to win it. And he did, not knowing, in his own words, how he managed to convince a company to pay a substantial sum of money to a newly formed startup in another part of the world that didn't have any real capability in the domain either!

This project provided some cushion for Prateep to hire the first three employees of SatSure – all from his alma mater IIST, who were graduating and did not get an offer from ISRO, having missed the CGPA (Cumulative Grade Point Average) cut-off. And that is how the founding team of SatSure was formed.

Abhishek managed to find an office space for cheap, which was literally above a garage and in a not-so-impressive neighbourhood of Bengaluru. Meanwhile, their results from the pilot project seemed acceptable to the Srikakulam district collector, and the team was trying hard to find budgets to scale it up. It was then that Prateep got connected to Hemendra Mathur, a partner at an IIMA-backed deep-tech fund and an agri-tech-evangelist himself. Having previously worked in agriculture finance at Rabobank and Yes Bank, Hemendra found the startup's pitch very interesting. He connected the team to a consulting firm that was preparing a report on agri-tech innovations in the country.

Upon speaking with this firm, the founders discovered they were supporting the Andhra Pradesh government in organizing an agri-tech summit in collaboration with the Bill & Melinda Gates Foundation. The event aimed to identify, nurture, and deploy cutting-edge AgTech (agri-tech) solutions across the agricultural value chain. Prateep filled out the application form at the last minute for participation in the summit's pitch competition which sought to identify the best AgTech solutions relevant to the challenges faced by small and marginal farmers in the state. He wanted to take his chances as it seemed like a potential place for finding budgets to scale up the pilot project in Srikakulam. And SatSure had a letter of endorsement from the district collector, so they fancied their chances of getting shortlisted.

As luck would have it, SatSure was shortlisted and emerged among the top three companies in the competition. After several months of discussions and deliberation with the senior bureaucrats at Andhra Pradesh's agriculture ministry, SatSure received a contract worth ₹5.13 crore. The bet had paid off, and this was the 'seed' funding the team required to take off.

The success at the Andhra Pradesh AgTech Summit acted as an endorsement of SatSure's idea by the government itself. It helped that Bill Gates flew down to felicitate the winners in person, so Prateep's picture with him and Chandrababu Naidu, the state chief minister, was all over social media and news channels. It piqued the interest of many because SatSure was legally only a few months old!

The startup started getting inbound interest from insurance companies, agribusinesses, VCs, and even Central ministry bureaucrats. The founders thought this could be the right moment to raise some venture funding too, so they spoke to a few known firms about their work. However, most of them had the same reaction – they were fascinated but unsure as to how a company like SatSure would make money without over-reliance on government support. Prateep had anticipated such reactions as VCs in India had mostly seen only e-commerce and SaaS (Software as a Service) companies succeed so far. Fintech was an upcoming trend back in 2018, and SatSure's work was at the intersection of agri-tech, space-tech, and fintech. However, the founders' determination and persuasive abilities played a pivotal role in the company diversifying its initial customer base by securing small but meaningful proofs of concept with insurance companies.

The journey from vision to execution was tough, as it had to happen very fast, and the founders were unprepared when it came to building teams and running a company. However, they had a never-say-die attitude and worked tirelessly on the ground, spending time with farmers to understand their needs and speaking to insurance companies, government, and banking customers to educate them about satellite imagery and its applications, while learning from them about the problems they faced in their jobs.

The success of the Andhra Pradesh project instilled confidence in SatSure among insurance companies, but like every other government engagement, it created a massive cash flow gap for SatSure. The small proofs of concept with insurance companies and banks were barely able to cover operating expenses like salaries and rent. Prateep was not drawing any salary and was living off his savings, so he started actively pursuing investments based on the traction they had generated.

A large Japanese conglomerate showed interest, and the conversation reached a point after months of discussions where a term sheet was issued at a very favourable valuation for SatSure. However, the conglomerate itself underwent a corporate reshuffle at the global level and the deal fell through, giving the founders a very valuable life lesson – not to put all their eggs in one basket. Everyone in the team was relying on the founders to pull this one off too, just as they had so many times earlier.

This setback made Prateep think deeply about a long-term, sustainable growth strategy. He knew that building a product could create more predictable revenues and his team at SatSure was still in the process of finding the right product–market fit. Meanwhile, their accomplishments were gaining recognition, with the World Bank naming SatSure as one of the top twenty insurtech companies, and the company also winning the MIT (Massachusetts Institute of Technology) Challenge, among other accolades. These achievements helped to boost their visibility within the Indian agri-tech ecosystem.

However, the increased attention also created a problem, as the small team found themselves inundated with a wide range of inbound queries. Prateep was aware that it would still be a few years before they could find out what the return on investment for a bank lending to farmers using SatSure's data insights would be. Until then, without a clear value-creation statement, he couldn't take the repeatability of the business for granted. There were technology risks involved, which his customers wouldn't be able to understand, especially with the noise created by some other agri-tech startups about the possibilities of using satellite remote-sensing data.

A safe bet during such uncertain times was to create government business-driven growth, but Prateep was invested in the vision they began SatSure with and was also afraid of the revenue concentration risks if he over-indexed on the government as a major revenue source. So, he took the risky call of co-building the product with a customer, convincing his co-founders and teammates about the merits of such an approach, especially in building customer credibility and later raising institutional capital.

ICICI Bank emerged as a potential customer. After conducting an extensive proof of concept with the bank to evaluate SatSure-generated data for lending to farmers, Prateep was told that while the results looked promising, ICICI's interest lay more in using the data for improving the efficiency of farm loan monitoring and collections. There was money on the table but not for the purpose for which Prateep and his team had originally anticipated.

Since the founder had already started refusing a lot of service-driven work, revenue visibility and business sustainability would be threatened if they refused to work on the proposed engagement by the bank. So Prateep bit the bullet and signed the contract, marking his first commercial sale for SatSure valued at more than ₹1 crore. Such a sudden requirement for scale (the requirement was for monitoring 70 per cent of the bank's portfolio, spanning 100 districts and 1,00,000 villages) was unanticipated by Prateep. Still, the team adapted quickly and had the backing of the customer too, allowing for mistakes in the initial period.

Prateep still remembers the closure meeting of this critical deal, as the negotiations had gone back and forth for a while, and he was down to his last ₹1,000 in his bank account. He had to borrow money from his sister to get home from the airport on his way back from Mumbai, but the risk proved to be worth it.

The founder knew they had to increase the team size at SatSure quickly, but instead of financing growth via equity, he started exploring debt – an unusual option for an early-stage startup. Being a startup with losses and less than three years of financial statements, getting a loan from a bank seemed impossible, and this was where SatSure found support from an upcoming non-banking financial company (NBFC) called Samunnati Finance, backed by Rabobank's foundation, the social fund of Rabobank that focuses on supporting farmers and their cooperatives in developing countries.

This allowed him to double the technology team size rapidly while continuing to work closely with their anchor tenant customer. This close engagement provided two significant benefits: first, it gave SatSure access to conduct deep research into the impact of their work on agricultural lending; second, it allowed them to validate and refine their product through actual users, expanding their capabilities beyond loan monitoring to potentially enable direct financing to farmers in the future.

Prateep spent a lot of time during this period in the field, and during a visit to a bank branch in Uttar Pradesh, he discovered that the branch manager was issuing a mere 20 farm loans per month, in stark contrast to the 200 or more other types of loans being processed. Upon inquiring about the reason for the low volume of agricultural loan sales, he learned that farmers were opting to seek money from private moneylenders rather than banks because of the convoluted, time-consuming, and complex loan processes at banks that involved land and ownership checks with the revenue circle officer, payment of legal fees for attestation of land ownership, and verification of the crop grown at the farm.

Prateep wondered whether he was trying to put the cart ahead of the horse by pushing a product that did alternative credit scoring of farms when the acquisition of basic data relating to land and ownership itself was broken. Was his space technology evangelist self not reading the signs that there were problems to solve for the banks before they entertained data from space for decision-making?

This made Prateep consider solutions to the digital land and ownership-check feature alongside creating satellite imagery-driven insights on farm revenue risk, leading to an increase in banks' capacity to lend more and lend faster. The turnaround time for crop loan approval ranged from 28 to 45 days, and that didn't come with any guarantee that the loan would be approved even after completion of all the required procedures. This stood in stark contrast to the ease with which personal loans could be obtained, where anyone with a simple CIBIL score could secure a loan within minutes.

Poor rural bank branch density added to these systemic inefficiencies, leading to farmers having to travel long distances to apply for loans. And they would often be turned away by the credit manager for trivial details found missing in the application process, leading to frustration on both sides. In summary, banks lacked the necessary means and tools to expedite the loan process, while farmers felt harassed and burdened by the cumbersome bank procedures. Prateep thought that failure to solve this problem could have a detrimental impact on this satellite and AI-driven farm loan enablement product idea.

After deep deliberation with his teammates, the founder understood that solving such a massive problem would require them to invest a lot more capital in the business, so raising equity capital was inevitable. As Prateep started readying himself to become a 'pitcher', March 2020 arrived, and along came the COVID-19 pandemic. The world came to a standstill, and within a span of a few days, his plans for fundraising changed to plans for the survival of the firm. No one knew how long the disruption to 'business as usual' could be, and rapidly, the world started falling apart around Prateep.

Customers became inaccessible. Employees were battling anxiety and adjusting to working from home. And investors, whom he had just started reaching out to, stopped responding to his emails or messages. Until then, Prateep had had the luxury of having well-wishers and senior advisers of experience guiding him. But the pandemic was a first-time experience of its kind shared by all generations – meaning, no one knew what to do!

Prateep, Abhishek, and Rashmit spent hours discussing and debating the fate of the company, but in all this chaos, something interesting started happening. Suddenly, there was a renewed interest from almost every potential customer that Prateep had either worked with earlier or spoken to in the past three years. Everyone believed that remote monitoring capability and data collection from space was a 'must have' and not just 'good to have' for their organizations.

While the general advice to entrepreneurs from their investors and advisers was to conserve cash and cut costs, Prateep's conviction in the strength of the business made him take the exact opposite approach. He called a meeting of all the key management within the company two months after the first lockdown and informed them that they would go in for a 'blitzkrieg' – a German concept from World War II, referring to their armed forces taking the enemy by surprise. Everyone thought he was crazy to say, 'let's go all in' while also acknowledging this could very well be 'all out', but Prateep had the good fortune of a team that believed in him.

To the surprise of most people who knew SatSure by then, the strategy actually worked. They onboarded more clients in the next four months than they had cumulatively over the last three years. The team size doubled within the same time frame, and all of this was happening without the cushion of venture capital in the bank. The first signs of arriving at a product–market fit finally started to show for Prateep.

But that excitement faded rapidly as he understood that with an expanding team, his leadership also had to evolve. For everyone, the times were tough on the personal front as burnout was creeping in, along with doubts as to whether SatSure could ever raise money. It wasn't an everyday story that a firm had scaled to $1 million in annual recurring revenues without equity funding. And it came at a cost for Prateep and his team – below-market salaries, no perks whatsoever, no work-life balance, and tonnes of anxiety about the future.

The founder spoke to some of his advisers about his plans to re-initiate a fundraising process. He consulted Hemendra Mathur, a venture partner at Bharat Innovation Fund and a noted agri-tech expert, and Arindom Datta, the Asia head of rural and development banking at Rabobank, known for his extensive experience in agricultural finance. Both advised that he look for strategic investors because agri-finance and space technology are niches in themselves, and SatSure operated at the intersection of both. However, the venture capital market in early 2021 was at the peak of the digital hype.

Prateep was already under the pressure of losing his team, which had stood by him since the beginning, but he was visibly stressed, financially and emotionally drained after the pandemic. He had imagined that raising capital would be easy because, in the past, he had been approached by several investors organically. Due to Prateep's refusal to raise capital before clear proof of his product thesis happened (a rare occurrence in the Indian startup ecosystem), a rumour had begun to circulate in the industry that SatSure's lack of external investment was due to poor business performance.

The founder grew anxious, wondering whether he had done the right thing by becoming an entrepreneur. He knew he had come a long way, from aspiring to a career in space technology to dealing with banks, investors, and the government in a sector (agriculture) that he had no clue about until four years back. Yet, he was unprepared for the mountains that had to be climbed, whether it was building an unconventional startup and bringing in revenues from the word go or raising capital while sticking to his core values.

Prateep's initial confidence in his ability to secure investment for SatSure, given that the company had been operating for four years and had revenues from a reputable customer base, slowly started to wane after repeated rejections by VCs. He found himself struggling to effectively tell the story of his company to potential investors, who cited various reasons for their refusal to fund it.

One investor expressed doubts about SatSure's scalability and remarked that the company's focus on the Indian market rather than the US market was a limitation. Another investor claimed that the total addressable market (TAM) for SatSure was only around $40–50 million. Prateep countered this argument, asserting that if they considered just one module of their product in agri-finance, the real TAM in India alone would exceed $100 million.

Yet another investor acknowledged SatSure's impressive performance but pointed out the absence of venture capitalists on the company's cap table and said they would not take the first bet. And yet another investor deemed it risky to invest in SatSure, saying the markets in which it operated were still in their early stages.

Prateep took this feedback in his stride, stopping to judge whether it was right or wrong because he was running out of time. He sought the help of a boutique investment banking firm based in Bengaluru, acknowledging that he needed help to raise capital. This was very different from selling to actual customers. After a few poor pitches, a couple of verbal beatings from Faheem (his friend from the investment bank) about his poor storytelling, and 36 rejections later, an unexpected opportunity presented itself when the founder received an inquiry from Baring Private Equity.

He looked them up and saw the logos of established companies on their website as part of their investment portfolio. Prateep had the lowest of hopes from this meeting, and he thought that perhaps this firm was trying to build its thesis in the area of his work through such outreach to entrepreneurs like him. He entered the meeting with low expectations, but to his surprise, it turned out to be the most engaging conversation he had experienced in a long time.

The investors showed little interest in assessing the technology being built; instead, they wished to speak with one of the company's customers before making an investment decision. Arul Mehra, a partner at the firm and leading the conversation, cheekily asked Prateep what any of his customers would say if he called them up. The founder, having gone into this meeting without any expectation, responded back in his witty way, saying they would tell Arul that SatSure is expensive, yet they still work with the company. And that, followed by the equity firm's call to a couple of the customers, led to the first cheque being written for SatSure!

Prateep's unwavering commitment to his original core value proposition of transforming agri-finance sets SatSure apart from other agri-tech startups. While many companies in the sector have diversified their offerings or pivoted to new business models, SatSure has remained laser-focused on addressing the pain points in agricultural lending. A key distinguishing factor is that Prateep and his founding team invested their own funds into the company. Each founding member brought a certain level of knowledge and capital to the table, with each holding direct stakes in the venture, their combined investment amounting to around ₹4.4 crore.

Maintaining a steady revenue stream was crucial to the survival of the business, so they implemented strict revenue discipline and ensured the timely collection of payments. They were prudent in their financial management, unlike other startups with access to substantial venture capital. They had also built a product with the farmer at the heart of the user experience, exploring ways to leverage satellite technology to improve every touchpoint of the loan acquisition process. This singular focus has been instrumental in guiding SatSure to where it is today.

To this end, Prateep and his team have devoted considerable effort to establishing a competitive edge in the analytics industry by collecting, understanding, and leveraging satellite data in innovative ways. They have invested countless hours analyzing vast datasets, such as thousands of satellite images, and developing algorithms to automate the analysis process. This has allowed SatSure to extract farm-level insights from data at a scale and speed that many competitors cannot match.

Despite their success in transforming hard technology into a socially impactful product, Prateep and his team remain committed to continuously improving its adoption and developing new offerings that cater to the needs of different types of farmers, including both landholding and tenant farmers. They recognize that there is still much work to be done to truly revolutionize the agriculture finance landscape and create a more inclusive and empathetic system.

Looking back, Prateep already feels that it has been a long journey, from being a rocket scientist to becoming part of the agri-tech community. The interlinked, incremental innovations done at SatSure by Prateep and his team through continuous engagement with real customers, whether it is the automation of satellite spectral data analysis or the development of user-centric designs for adoption in the rural areas, have helped SatSure differentiate itself in an increasingly crowded market. By providing their clients with unparalleled insights, ease of use, and tangible value, Prateep and his team have established a strong competitive advantage, which has been a key contributor to the company's success.

Today, SatSure is a company that employs more than 150 people and has offices on three continents. Post the investment by Baring Private Equity, some of their customers such as ICICI Bank and HDFC Bank also ended up investing in SatSure and becoming supporters of the vision Prateep and his team set out on – to transform how farmers have access to finance using data from space. The company recently launched a financial product in partnership with CIBIL and has begun exploring opportunities in other sectors, such as seeds and fertilizers, where their proprietary algorithms and patented technology can be utilized to improve seed quality and address regional issues. They are also actively working to support the agricultural supply chain, recognizing the importance of a holistic approach to transforming the industry.

As SatSure continues to grow and evolve, Prateep and his team remain committed to their moral and ethical responsibility to society. They firmly believe that technology and data-driven insights have the power to drive positive change in the agricultural sector, empowering farmers and promoting financial inclusion. Through their innovative solutions, dedication, and compassionate approach, SatSure is reshaping the landscape of agri-finance and paving the way for a more sustainable and equitable future for farmers across India.

In this journey, Prateep's personal story serves as a testament to the transformative power of conviction, resilience, and determination in the pursuit of one's passions. From his humble beginnings in Ranchi to his meteoric rise as the founder of one of the country's most well-known startups, Prateep's journey is a source of inspiration for aspiring entrepreneurs and change-makers alike. His story reminds us that with hard work, perseverance, and a steadfast commitment to one's values, it is possible to overcome even the most daunting challenges and make a lasting impact in the world.

# 3

# The Tinkerer Turned Titan: Karthik Jayaraman, WayCool

*F*ROM DISASSEMBLING TOY CARS *as a child to revolutionizing India's agricultural supply chain, Karthik's journey is a testament to the power of curiosity,* unconventional thinking, and relentless problem-solving. His life has been a series of calculated risks and bold choices that often defied conventional wisdom. Whether opting for mechanical engineering over the more lucrative computer science, choosing Purdue University for its hands-on approach, or returning to India to join Tata Motors when his peers sought their fortunes abroad, Karthik consistently charted his own course. This maverick spirit, coupled with his insatiable appetite for learning, led him to pursue an MBA at the Indian School of Business (ISB), Hyderabad, when he recognized the need to hone his leadership skills. Karthik's strategic thinking, sharpened by his diverse experiences and his work at continuous self-improvement, ultimately culminated in the creation of WayCool – a company that embodies his lifelong passion for mechanics, innovation, and social impact. In Karthik's story, we see how a childhood fascination with tinkering can evolve into a transformative vision, reshaping an entire industry and improving millions of lives along the way.*

In the thriving southern city of Chennai, a young boy's passion for mechanics was ignited by a simple gift – a mechanical toy car. As his tiny hands explored its intricate workings, Karthik's curiosity blossomed. Days flew by as he immersed himself in the world of toy automobiles, tinkering with them and repairing them whenever they broke down.

At the age of eleven, the young enthusiast's world expanded when he received a Lego kit for his birthday. Demonstrating an innate aptitude for problem-solving, he constructed over a hundred mobile vehicles, showcasing his creativity and ingenuity. These early experiences laid the foundation for his future in the automotive industry, developing in him a keen eye for detail and an understanding of complex mechanical systems.

As Karthik entered ninth grade, his parents began discussing his future, suggesting engineering as a career path based on his interests. The family's hopes crystallized when they found that his cousin had been accepted into IIT Madras, one of India's most prestigious engineering institutions. This introduction to IIT fascinated the young boy, but his parents had to explain the complications to him: IIT with its merit-based admission and subsidized education was one of the few options for providing him with a top-tier engineering education, given the high fees of other colleges and limited seats. In other words, it was not just the best choice but also the only choice for him – and an incredibly difficult one to get into.

This revelation shocked the aspiring engineer, who suddenly realized that his dreams hinged on securing a spot at IIT. The pressure was immense, but he rose to the challenge. For the next three years, his life revolved around preparing for the JEE, which had a daunting success rate of less than 2 per cent. But Karthik dedicated himself entirely to his studies. Every day, he cycled to his coaching class, studied late into the night, and woke up early to revise. The routine was gruelling, but the young man remained committed, driven by his family's hopes and sacrifices.

Just a few months before the exam, a setback struck when he fractured his leg in a sports accident. The injury added to the mounting pressure, but it also strengthened his resolve. His parents rallied around him, ensuring he didn't miss a single day of preparation despite his condition.

This challenging period brought about a profound shift in Karthik's perspective. He realized he wasn't just pursuing his dream for himself but for his family too. The desire to make them proud and repay their sacrifices with his success tapped into reserves of energy and focus he didn't know he had, pushing him to achieve things he never thought possible.

When the results of the JEE were announced, it was evident that Karthik's hard work had paid off. He secured the 70th rank nationally, earning him a spot at IIT Madras. It was a moment of pure joy and relief for him and his family.

As he prepared to start his studies at IIT, he faced pressure from peers and relatives to reconsider his choice of major. Many urged him to pursue computer science, seen as the most lucrative and in-demand field. Despite briefly enjoying coding, the young man knew his true passion lay in mechanical engineering. He wanted to work with his hands, designing and building machines to solve real-world problems.

At IIT Madras, Karthik faced his first academic struggles, particularly in physics. Undaunted, he redoubled his efforts, seeking help from professors and classmates. As he delved deeper into subjects like applied mechanics and fluid dynamics, his interest in and understanding of them grew. The hands-on aspects of his coursework, like operating machine tools and shaping materials, foundry, and forging, gave him a deep sense of fulfilment.

As he progressed through his degree, the budding engineer became increasingly convinced that mechanical engineering was the right path for him. 'If you don't mind working with your hands, mechanical engineering is a damn good field,' he would often say to his friends.

Approaching his final year at IIT, Karthik focused his job search on automobile manufacturers like Tata Motors, Mahindra, and Maruti Suzuki. To his delight, he received an offer from Tata Motors, one of India's most prestigious automobile companies. The prospect of contributing to the development of new vehicles thrilled him.

Yet, before he could fully savour his success, a new challenge emerged. His parents, relatives, and friends began urging him to pursue a postgraduate degree at a foreign university instead of joining Tata Motors. They argued that a degree from a prestigious institution like MIT or Stanford would open up even more opportunities for him in the future.

The young engineer found himself torn. On one hand, he appreciated the opportunity to study abroad and expand his horizons. On the other, he felt a deep sense of loyalty to his country and family. After all, it was the Indian education system that had nurtured his talent and enabled him to reach this point. He felt indebted to contribute to the growth of the Indian automobile industry.

Despite his reservations, Karthik eventually yielded to the persistent arguments of his family. He agreed to go abroad for a short time, promising to return to India after graduation. He began preparing for the GRE and TOEFL exams, aiming to secure admission to a top university.

His efforts bore fruit, and he received offers from several prestigious institutions, including Purdue University and MIT. While everyone advised him to choose MIT, Karthik had his reservations. His research revealed that MIT's focus was on computer-aided design/manufacturing, which didn't align with his interests as much as hands-on mechanical engineering. In contrast, Purdue University's strengths lay in manufacturing science and technology, perfectly matching his passions.

Defying the pressure from those around him, the young man decided to follow his instincts and enrol at Purdue. It was an unpopular choice, but he knew he had to stay true to himself and his aspirations. The presence of many of his friends from IIT at Purdue would provide an additional support system for him in a new country.

At Purdue, Karthik immersed himself in his studies, taking full advantage of the university's excellent facilities and hands-on approach to engineering. He conducted experiments, worked on challenging projects, and expanded his knowledge of manufacturing science and technology. This experience boosted his confidence and taught him the importance of autonomy and innovation.

As graduation approached, the aspiring engineer faced a crucial decision. While his peers were accepting lucrative offers from US companies, his heart was set on returning to India. His timing, however, couldn't have been worse. India was facing an industrial recession in the middle of 1998, and hiring had slowed significantly.

But Karthik remained optimistic and proactive. He reached out to various automobile manufacturers in India, showcasing his skills and passion for the industry. His persistence paid off when he received an offer from Tata Motors, his dream company. Despite having other options, including offers from Bajaj Auto and Mahindra, he chose Tata Motors, drawn by their reputation for innovation and their recent success with the 'Made in India' Tata Indica car.

At Tata Motors, Karthik joined the cabin design team, where he had the opportunity to learn from highly skilled engineers and collaborate across various departments. One of his key learnings came from interactions with the production team. Initially tasked with relaying problems between departments, the young engineer took the initiative to understand the production team's needs and devise solutions himself. This proactive approach impressed the production executives and taught him the importance of being a problem-solver rather than just a messenger.

Despite the valuable experience he was gaining, Karthik grew increasingly frustrated with the slow pace of change and the constraints of working within a large corporate structure. The bureaucracy and hierarchy often stifled innovation and made it difficult to implement new ideas quickly. He yearned for more autonomy and the ability to rapidly implement innovative solutions.

During this period of reflection, an unexpected opportunity arose. Timken, an American bearings manufacturing company, reached out to Karthik, seeking his help in establishing their new R&D centre in India. Impressed by Timken's vision and culture, and seeing a chance to make a greater impact, he made the difficult decision to leave Tata Motors.

At Timken, the young engineer thrived in an environment that valued individual empowerment and innovation. His ingenuity shone through when he developed a cost-effective, locally manufactured solution for a faulty machine component, saving the company thousands of dollars and weeks of downtime. This experience sparked an idea that led Karthik and his team to manufacture some of Timken's test equipment in India at a lower cost than previously, which was eventually deployed across various Timken factories worldwide. This achievement not only demonstrated his ability to think outside the box but also highlighted the potential for cost-effective, locally manufactured solutions.

As Timken India grew, so did Karthik's responsibilities. He found himself making critical non-technical business decisions, including those related to people management, resource allocation, and strategic planning. Working under multiple supervisors with different perspectives on problem-solving, he initially struggled to make the right decisions at the right time. However, one of his supervisors, who had studied for an MBA and worked as a consultant at McKinsey, gave him valuable advice on making and evaluating business and management decisions. This experience ignited Karthik's interest in pursuing an MBA, and in 2003, he applied to and was accepted into ISB.

The decision to attend ISB proved to be a wise one. Unlike the IIMs, where most students were recent graduates, ISB attracted experienced professionals from diverse backgrounds. Karthik's classmates were well-established entrepreneurs and senior executives running large organizations, and he found himself learning as much from them as he did from his coursework. The diverse perspectives and opinions shared by his peers were invaluable insights into business, management, leadership, and networking.

Upon completing his MBA, Karthik received job offers from his former employer (Tata Motors), Timken, and McKinsey. His passion for automobiles naturally drew him towards the automotive job, but his friends convinced him to consider the McKinsey offer more seriously. They argued that he could always join an automotive company later, but the opportunity to work for McKinsey was a once-in-a-lifetime chance. Realizing that a stint at the prestigious consulting firm would accelerate his learning of business and management skills, he accepted the offer from McKinsey, a decision that would prove crucial to his future success.

At McKinsey, Karthik initially struggled to find his footing. His extensive work experience and industry knowledge set him apart from his younger managers and colleagues, who were mostly fresh graduates from IIT and IIM. This difference made it challenging for him to relate to his peers, and he found himself becoming isolated. Every day, he contemplated leaving the company, but he resolved to stay and complete the first project he was assigned.

The project, focused on organizational transformation, became a turning point for the young consultant. As it moved into the execution phase, he found his stride. Karthik's approach was unique – instead of imposing solutions, he worked collaboratively with employees and managers, implementing small experiments and prototypes to solve problems incrementally. His ability to listen to and understand stakeholder concerns allowed him to adjust solutions for successful implementation.

The project's success was remarkable, with the client company growing from ₹600 crore to ₹3,200 crore in revenues. Karthik's two-year involvement made it the longest project ever undertaken by a McKinsey associate, leading to several promotions for him and accelerating his career.

In his next project with a cement manufacturer, Karthik's innovative approach shone again. Responding to an employee's concern about career stagnation at the cement company, he and his team proposed a bold solution: restructuring positions based on employee strengths and weaknesses. This approach aimed to improve individuals' self-awareness, emotional intelligence, and skills. The impact was profound – four years later, the same employee who raised the initial concern had become the CEO of a multinational company, showcasing the power of Karthik's people-first approach.

Despite his success at McKinsey, the desire to be closer to his ageing parents and newborn daughter led Karthik to join Ashok Leyland in Chennai. Here, he faced new challenges in creating a corporate strategy to manage emerging competition from global brands. He established a clear vision with a five-year plan, breaking it down into manageable yearly goals and implementing a feedback framework. However, the company struggled to achieve the vision.

Intrigued by this setback, Karthik investigated the root cause and discovered that the different divisions of the company operated in silos, primarily due to a lack of cross-functional skills among the leaders of those departments. This had been the company culture. His solution was the innovative 'emerging leaders programme', which identified top employees and assigned them critical cross-functional projects. The programme's success uncovered hidden talent within the organization and became a model for future initiatives. The identified leaders were put on an accelerated growth trajectory and tasked with leading divisions, regions, or even the entire company in the future.

Karthik's leadership was put to the test during the 2013–14 auto industry recession, which severely impacted Ashok Leyland. The sudden drop in market demand left 13,000 Leyland trucks stranded, and the company was left with only 88 days of working capital. Karthik and his team, under the leadership of Vinod Dasari, quickly transformed the emerging leadership programme into a company turnaround programme, initiating projects such as a cost-reduction programme and raw materials reduction programme. However, Karthik made it clear that these programmes would only cut excess fat, not muscle.

The results were impressive: working capital was reduced from over 88 days to a negative 14 days, saving over ₹1,400 crore through modularization and other tools – an unprecedented feat in the mature automotive industry. The turnaround efforts transformed Ashok Leyland from a regional to a pan-Indian company, significantly increasing its market share. However, the intense work left Karthik exhausted.

Moreover, Karthik observed that the automobile industry was on the verge of decline as a result of the rise of electric vehicles, the replacement of driver-driven cars with vehicles that had self-driving capabilities, and the replacement of owned cars with rental cars. He believed that the engineering design of electric vehicles was simpler than that of cars with internal combustion engines, and the advent of autonomous vehicles raised questions about the driving experience designers could offer.

This realization, coupled with his exhaustion, prompted Karthik to make a bold decision. Despite his success and years of experience in the automobile industry, he chose to leave it and explore new opportunities. It was time for a new, stimulating journey that would leverage his diverse experiences and innovative approach to problem-solving.

After departing from Ashok Leyland, uncertainty clouded Karthik's next steps. He contemplated various ideas, including acquiring a defence firm and launching a company to manufacture mechanical components. During this period of reflection, Sanjay Dasari, son of one of Karthik's close friends, became a frequent visitor, sharing his business ideas with him. Sanjay, having already founded two companies from scratch, was eager for a new venture.

One day, during a casual conversation, Sanjay pitched his latest concept: Korean BBQ food trucks catering to Chennai's Korean population. As they delved deeper into the idea, they uncovered a web of issues plaguing the food supply chain – from hygiene concerns to unstable supply and unfair compensation for farmers. This revelation sparked something in Karthik, whose expertise in logistics and supply chain management suddenly found a new purpose.

Intrigued by the challenge, Karthik conducted a preliminary analysis. The numbers showed promise, and with Sanjay's father, Vinod Dasari, offering to be their first angel investor, the seeds of a new venture were sown.

Cautious by nature, Karthik suggested that they needed to understand the supply chain thoroughly before proceeding. What followed was a journey that would transform their perspective forever.

The partners embarked on a mission, tracing the path of fruits and vegetables from plate to farm. Their investigation led them to Koyembedu, Chennai's largest produce market, where they unravelled a complex tapestry of challenges facing India's agricultural sector.

The research revealed several alarming facts. First, the supply chain was convoluted, lengthy, and filled with numerous intermediaries. This complexity stemmed from the small size of Indian farms, which averaged only 2–2.5 per cent of a typical 440-acre American farm. Moreover, Indian farmers produced only 500 kg of a particular vegetable per farm, compared with the ten trucks produced by American farmers from a plot of the same size. This necessitated the aggregation of produce from multiple small farms, and the non-uniform produce from these different farms required further intermediaries to grade the produce into different categories, each sold to different customers at varying prices.

The second challenge lay in the purchasing habits of Indian consumers. Most did not buy groceries from supermarkets, instead they procured 90 per cent of their fruits and vegetables from small, local kirana shops on a weekly basis. This practice, which allowed for quick, nearby purchases, was unlikely to change in the near future. As a result, kirana stores served small areas and required modest supplies of produce. This meant that large trucks delivering 25 tonnes of vegetables to wholesale markets like Koyembedu had to divide their stock to supply these smaller stores.

Complicating matters further, traders in the wholesale markets didn't sell directly to kirana stores. Instead, they auctioned their stock to semi-wholesalers, who then supplied the smaller shops. This system created a need for multiple intermediaries – aggregators, small traders, and mandi auctioneers – to keep the supply chain operational. While these individuals charged for their services, leading to higher prices for produce, most earned only between 3 per cent and 10 per cent profits.

A third challenge arose from the multiple handling of produce as it moved through the supply chain. Frequent transfers between containers often damaged the goods and increased final consumer prices. Additionally, stakeholders in the supply chain incurred substantial losses when produce wasn't handled carefully. These losses added to the total cost, further inflating prices for consumers.

Another significant issue was the farmers' inability to predict future demand. Relying on often inaccurate and unreliable information from local markets, they created fluctuating supply despite relatively constant demand. This inconsistency led to significant losses for all supply chain participants and excessive pricing for end consumers.

For instance, if tomato prices rose in a nearby market, farmers might assume increased demand and plant more tomatoes. This would spark higher demand for tomato seeds, prompting local seed merchants to increase their stocks. More farmers would then purchase tomato seeds to avoid missing out on a potential tomato boom, ultimately leading to an oversupply of tomatoes. When farmers couldn't find buyers, tomato prices would crash, causing severe losses to them. This phenomenon partly explains why farmers sometimes resort to discarding produce, throwing their produce onto roadsides or into rivers.

In India, demand for vegetables and fruits remains relatively stable. Recognizing this, Karthik and Sanjay realized that aligning the supply chain with demand would be beneficial in preventing losses and maintaining stable consumer prices. This insight formed the core strategy of their new company, WayCool – a demand-driven supply chain model rather than a supply-driven model.

Armed with this knowledge and a clear vision, the two set out to build WayCool, aiming to revolutionize India's agricultural supply chain. Karthik's extensive experience in logistics and supply chain management, combined with Sanjay's entrepreneurial spirit and Vinod's support, provided a strong foundation for their venture.

As they began implementing their plan, they quickly realized their biggest challenge would be to convince both customers and suppliers to trust their vision. Initially, they approached professionals in the hotel industry, hoping to secure orders and establish a reliable customer base. However, they were met with hesitation and scepticism. Hoteliers proved reluctant to take risks and switch from their current vendors, forcing the founders to pivot their strategy.

The founders then decided to become retailers themselves, selling directly to end customers. Drawing on Sanjay's experience with mobile food trucks and Karthik's engineering skills, they created a store on wheels, a novel concept for Chennai's residential areas.

On their inaugural day, they drove to a residential area and parked the truck, eagerly awaiting their first customers. Despite the concept being unique, the initial response was underwhelming. However, Sanjay's keen eye spotted an opportunity. Noticing someone selling lemon juice nearby, he purchased their entire stock, offering it for free to passers-by. As people stopped to enjoy the refreshing drink, curiosity about the mobile store grew. To their surprise, they discovered an array of high-quality fruits and vegetables inside the truck. Soon, purchases were being made, and word about it spread throughout the neighbourhood. Before long, WayCool's entire stock was sold out.

Encouraged by this success, the team now faced the challenge of securing a reliable supply of produce from farmers. Fortunately, one of Karthik's friends, who managed a well-known dairy brand, had established connections with farmers and helped secure an initial supply. With produce in hand, the next hurdle was finding a suitable place to operate from and store their products. They leased a modest 600-square-foot space on the ground floor of a building in Chennai and began grading, sorting, and packing the fruits and vegetables before loading them onto trucks for delivery.

The company's innovative approach quickly gained traction. Within a few months, they had sold produce to more than twenty-five large residential complexes, expanding their fleet from one truck to seven. Although profitability remained elusive, the founders were gaining invaluable experience and refining their business model.

As they contemplated their next phase, the duo experimented with fixed retail stores. Their first attempt, a small, 100-square-foot store in Thorapakkam, Chennai, failed to attract customers as it operated out of a cramped space. They tried again with a slightly larger store in an underserved location, and this gradually gained traction.

Just as they were finding their footing in the retail space, an unexpected opportunity arose. A friend informed Karthik about a large residential complex in Mahindra World City, 60 km outside Chennai, struggling to receive fresh vegetables. Despite their initial hesitation to serve this locality as it was at a considerable distance from their warehouse, Karthik decided to take a leap of faith and open a store in the complex. Ten years later, that store remains a thriving part of their retail presence, marking a pivotal moment in the company's history.

As WayCool expanded, they encountered a well-designed mobile truck selling produce in DLF Garden City, Chennai.

Upon investigation, Karthik discovered that the truck belonged to a team sharing his vision of building a farm-to-fork supply chain. Despite their passion, this team struggled to secure funding for growth. Recognizing potential synergy, the founders proposed joining forces. One member of the team, Senthil, an organic farmer with extensive knowledge of fresh procurement and supply chain management, joined WayCool to lead those departments. This merger expanded the company's fleet to eight trucks and increased its retail presence to five outlets. The increased scale and visibility began to establish WayCool as a significant player, earning farmers' trust and respect.

Meanwhile, in Tindivanam, Tamil Nadu, an Indian multinational IT company's non-profit organization supporting local farmers was facing difficulties in selling their produce. Karthik reached out to them and established a partnership to purchase their products. He also collaborated with the National Agro Foundation, a non-profit organization focused on rural development and sustainable agriculture, which connected the company to several farmer groups and streamlined their sourcing process. As their network grew, the founders made the strategic decision to gradually sell off their trucks and focus on the supply chain.

With a reliable supply of produce secured, Karthik and his team faced a new challenge: farmers in and around Tindivanam, Tamil Nadu, who were now dedicatedly producing for the company, were collectively yielding more than 300 kg of fruits and vegetables daily, while demand was only 100 kg. To address this surplus, the team proposed to deliver the excess supply directly to local hotels. Hotel managers, previously limited to grade-B vegetables, were thrilled to receive grade-A produce from WayCool. This pivot allowed the business to scale rapidly.

Seeking to fuel further expansion, the founders approached venture capitalists for investments. However, they were mostly met with scepticism. Investors argued that physical retail in India wasn't scalable and that foreign direct investment in retail was prohibited, making it an unattractive prospect. One investor, however, suggested that the company focus on expanding its B2B presence. This advice prompted another pivot: moving retail stores into a subsidiary and shifting the sales focus to larger retailers. Leveraging their network, they connected with chefs at the larger hotels, who were impressed with WayCool's product quality and began placing orders. As these chefs moved between hotels, they took the business with them, further expanding the company's reach.

To accommodate this rapid growth, the company relocated to a massive warehouse previously used for leather manufacturing. Drawing on their automotive industry experience, the team implemented various mechanical and technical innovations to automate processes. Karthik firmly believed that reliance on manual labour was neither scalable nor cost-effective, potentially leading to poor productivity, conflicts, and loss of quality control. The modern, automated warehouse not only streamlined operations but also attracted new customers, leading to a surge in orders.

As the business continued to expand and innovate, new opportunities emerged. A customer requested vegetable supplies for their resorts in mountainous regions like Yercaud, the Nilgiris, Ooty, and Kodaikanal. Already transporting produce from these areas to Chennai, Karthik saw an opportunity for reverse logistics by utilizing the empty trucks returning to these locations. This innovative approach optimized transportation resources and allowed them to serve a wider customer base.

Word of their efficient supply chain management spread quickly. A large Bengaluru-based catering company sought their assistance in procuring quality produce. The WayCool team assigned a leader to relocate to Bengaluru and establish a new supply chain there.

Once the Bengaluru centre became operational, an interesting pattern emerged. Large quantities of vegetables were being transported between Bengaluru and Chennai in both directions. Recognizing an opportunity to maximize efficiency and reduce costs, the team integrated both transportation sides, using specialized trucks to ensure no vehicles ran empty. This approach not only saved them substantial money but also allowed them to offer lower prices to customers, fuelling rapid business expansion.

By this point, most customers realized that the company was more than a trader – it was an engineering firm running a sophisticated operation capable of efficiently moving materials over 400–500 km daily. This realization attracted investor attention, and they successfully secured their first investment of $2.5 million, positioning them for further growth.

As expansion plans took shape, a larger warehouse was needed. They discovered an old government building in Virugambakkam, Chennai, which met their requirements, albeit larger than initially planned. Despite the risk, Karthik decided to seize the opportunity, investing heavily in advanced technology and facilities that far surpassed those of their previous warehouse.

While some experts criticized this substantial investment in a single warehouse, Karthik saw it as a market enabler. The impressive facility became the company's best marketing tool, helping to persuade several large customers to partner with them.

Despite their growing success, a new challenge emerged with their primary customers – hotels and restaurants often struggled to pay on time, straining cash flows. To mitigate this risk, the team considered diversifying their customer base by supplying well-paying large retail chains. They hired an experienced retail buyer to secure orders, but progress was slow.

Determined to understand the retail industry better, Karthik began engaging directly with retail chain representatives. He discovered that retailers would only place orders with companies already established in the retail industry.

Around this time, they learned of a Bengaluru-based firm specializing in importing, packing, and selling fruits to large retailers. With its promoters looking to exit, Karthik saw a strategic opportunity and decided to acquire the business. This acquisition proved transformative, as the firm was well-known for supplying several modern retail chains. Moreover, it imported fruits through the Chennai port, aligning perfectly with the existing logistics network.

Following the acquisition of this firm, WayCool began supplying fruits to retail chains. Leveraging their efficient supply chain and network, they delivered well-packed, consistently high-quality fruits at competitive prices and on time. Impressed retailers began to trust the company, leading to orders for other fruits and vegetables.

While the retail business expanded, new challenges emerged. The subjective nature of produce quality led to lengthy payment delays from retailers until quality issues were resolved. These delays increased working capital requirements, straining finances. Additionally, retail chains drove hard bargains on prices, squeezing profit margins.

Recognizing these issues could potentially jeopardize the company if left unchecked, Karthik with the guidance of Chinna Pardhasaradhy, his CFO and long-time colleague, devised a two-pronged solution. First, he decided to start supplying small retail customers to diversify their base and reduce dependence on large chains. Second, Karthik proposed leveraging their existing supply chain by adding more product categories, such as pulses and rice.

A team member took charge of building the business of supplying produce to small kirana stores. Despite numerous challenges, he succeeded in establishing a strong presence in this market segment. Concurrently, a colleague from Andhra Pradesh, experienced in the rice industry, proposed the sale of rice through the company's supply chain and retailer network.

This proposal led Karthik to an epiphany. He realized that WayCool had been organically growing their fruit and vegetables business, starting with retail customers, then small hotels, and finally large hotels. While successful, this method was expensive and time-consuming. For rice, Karthik envisioned starting with bulk sales. Although it required significant investment and carried some risk, he believed that selling in bulk would help build scale in buying, which could then facilitate entry into the retail market.

A colleague, Chinna, knew a trader supplying rice in bulk to Sri Lanka. They approached him, offering to supply rice at a lower price than he was paying currently, with superior quality, less wastage during transit, and prompt delivery. Initially sceptical, the trader was intrigued by their confident pitch. After some negotiation, he agreed to a trial order of one container.

Despite the many challenges, the team successfully fulfilled the initial order. Impressed, the trader placed four more container orders, which soon escalated to eight, then forty. As orders reached forty containers, the company found itself at a critical juncture. Fulfilling the order required investing all their available funds – ₹2 crore – to purchase the rice. It was a high-stakes gamble; if the trader failed to pay them, they would be in dire straits. Taking the calculated risk, they proceeded, leaving just ₹4 in their account. Fortunately, the trader paid them promptly, allowing operations to continue. This experience not only helped them achieve scale in rice procurement but also enhanced their supply chain efficiency in transporting rice.

Buoyed by their success in bulk rice sales, Karthik formed a small team to approach retailers, offering 25 kg rice bags. Initially, they supplied rice in the bags provided by the millers, which often varied in colour depending on availability.

For a while the sales progressed smoothly. However, one day, salespeople reported an unexpected problem: customers were suddenly rejecting their bags of rice. Concerned, Karthik personally investigated the issue. He discovered that the millers had recently switched from orange bags to green ones, and customers were expressing doubts about the quality of rice in the new packaging.

This incident proved to be a turning point. Karthik realized that the orange bags had inadvertently become a proxy for quality in customers' minds. Despite the rice being identical, the change in the colour of the bag created a perception of inferior quality. The experience illuminated the critical importance of branding in building customer trust and loyalty.

Acting swiftly, Karthik and his team created their own brand called Madhuram, complete with a distinctive identity and consistent packaging. From that point forward, they sold rice under this new brand name. The rebranding reassured customers and boosted sales, marking another milestone in the company's journey.

As WayCool solidified its market position, new challenges emerged. In 2018, several startups entered the grocery delivery industry, offering discounted products that impacted sales. Karthik, however, remained committed to profitability and refused to engage in a price war. Instead, he and his team strategized on defending their market share, identifying two key approaches: enhancing product branding and leveraging their existing supply chain to acquire more customers.

Their extensive network of warehouses and supply chains, connecting them to numerous small kirana stores, presented a unique opportunity. The team pondered who else could benefit from this robust distribution network, and the answer became clear: FMCG (Fast Moving Consumer Goods) companies. They began distributing FMCG products alongside their own offerings to kirana shops, a strategy that quickly proved successful.

Soon afterwards, Pepsi approached them to distribute their products, marking a crucial turning point for the company. Being recognized as a Pepsi distributor opened doors, facilitating the expansion of their store network and enabling the cross-selling of rice and other staples.

A former Pepsi regional manager, who had recently moved to Marico, recognized WayCool's potential and invited them to distribute Marico products in Bengaluru. This new partnership required a sophisticated supply system covering 1,850 stores. Rising to the challenge, Karthik and his team modified their supply chain and streamlined operations using technology to meet Marico's requirements. Their efficiency and adaptability garnered industry attention, attracting other major brands such as ITC, Nestlé, and HUL. Each new partnership allowed them to expand their retail store network and increase opportunities to sell their own products.

As the company grew, team members began identifying new expansion opportunities. An accounts department colleague with good contacts in Viruthu Nagar, Tamil Nadu – a region known for its production of dals – proposed the establishment of a dal supply chain. Recognizing the potential, Karthik provided him with the necessary resources, and the colleague successfully built and continues to operate the dal supply chain efficiently.

Another team member saw an opportunity in the dairy sector and developed a dairy supply chain. WayCool began selling packaged milk under the brand name Shuddha. However, the company struggled to make significant profits from milk sales, as they were merely buying milk from a vendor, packing it, and selling it without adding value for the customer. Moreover, they faced stiff competition from established giants in the milk industry.

To overcome these challenges, the team leased a dairy, hired a dairy management and technology expert, and began sourcing milk directly from farmers. This move improved their margins, but the business remained unprofitable. They then decided to develop value-added products from milk to boost profitability. Market research revealed a growing demand for curd in the cities, particularly among households with working spouses who lacked the time to make curd at home. WayCool launched a curd brand in Chennai, which saw immediate success and earned good margins as a value-added product.

Around March 2019, an investor who had previously rejected WayCool was looking to divest his stake in Freshey's, a Chennai-based startup manufacturing idli/dosa batter. Karthik saw this as a strategic opportunity. He believed that occupying more space in customers' minds and refrigerators would make it easier to build a sustainable business and compete against emerging competition. With milk and curd already in the product portfolio, dosa/idli batter seemed a natural fit. Additionally, Freshey's boasted an excellent team, a highly automated facility, and access to 1,200 stores. Furthermore, WayCool could leverage its large volumes of rice and dal purchases to make the batter.

The acquisition proved highly successful. By 2022, Freshey's expanded from 1,200 to 8,000 outlets, becoming India's second-largest batter brand. Pre-acquisition, Freshey's was losing ₹40 lakh monthly, with ₹20 lakh in revenue; post-acquisition, it began earning ₹1.6 crore per month in revenue. This growth trajectory continued, with monthly revenue reaching ₹6 crore by 2024, making it a profitable business unit.

Freshey's also introduced paneer, another value-added refrigerated product, and eventually discontinued milk sales because of its low profitability. The brand expanded its presence beyond Chennai, making its products available across Tamil Nadu and Kerala. In the Chennai region, WayCool's paneer and curd products have already reached the number three position in market share, further solidifying the company's success in value-added dairy products.

By the end of 2019, the company was operating smoothly, but the COVID-19 pandemic in early 2020 presented them with an unprecedented challenge. While they had anticipated potential disruptions and implemented additional hygiene and social-distancing measures, the nationwide lockdown brought new hurdles. The leadership team considered temporarily closing the business but was concerned about investor reactions. To their relief, when they approached the board and investors, they found alignment in prioritizing people's safety. However, the team insisted on continuing operations, citing the essential nature of their products and the need to protect farmers' livelihoods.

They swiftly adapted to the changing COVID-19 landscape, implementing several key changes. Non-essential employees shifted to remote work, while part of a warehouse was converted into a dormitory for essential workers. The team arranged emergency travel passes and reorganized supply routes to focus on intra-state sourcing and delivery. Additionally, the company expanded its infrastructure, building 13 new warehouses and relocating grading and sorting facilities closer to farmers, reducing their need to travel.

Concurrently, Karthik and his team established a technology division in Bengaluru, which developed over 50 apps to digitize the supply chain and implement SAP (Systems, Applications, and Products in Data Processing) across all organizational functions. These efforts ensured continued product supply, lowered end prices for customers, and made the supply chain more efficient and resilient to disruptions. They also expanded to several cities and established a presence in tier 2 cities in South India through franchises. These initiatives contributed significantly to the company's growth, with projections indicating over ₹1,600 crore in revenue for the year 2022.

As a demand-driven supply chain, WayCool adopted a consumer-driven model to predict demand with over 80 per cent accuracy. Orders are collected through a mobile app, and data on purchasing patterns and customer behaviour is analyzed to determine optimal target prices for products, factoring in transportation, grading, and processing costs. Farmers are informed of predicted demand and prices; if they are satisfied with the offer, the company pays them upfront and collects the produce. This approach enables farmers to plan their harvests more effectively and reduces their risk of being left with unsold produce. Collection centres are located closer to farms, saving farmers' time and money on transportation.

The company has invested heavily in technology throughout its supply chain, including an inventory management system that maintains inventory levels of less than 1.5 days and tracks stocks in real time. By keeping supply slightly below demand, WayCool reduces inventory and waste, resulting in lower wastage than the industry average. Despite this focus on efficiency, the inventory management system is designed to maintain a positive customer experience, demonstrating that a company can be both efficient and customer-focused with the right technology and processes. These efforts have streamlined operations, reduced waste, and improved both the quality and pricing of WayCool's products for customers.

As it solidified its position in the South Indian market, the company began stretching its roots beyond its southern stronghold, venturing into the bustling markets of Maharashtra and even making its international debut by supplying to the Middle East. This geographic expansion was accompanied by product diversification, sourcing 23 types of fruits and vegetables from 15 countries worldwide.

To fuel this rapid growth and diversification, Karthik made a decision that seemed logical at the time – bringing in experienced professionals from outside the company. These new hires, with their impressive resumes and fresh perspectives, were given significant autonomy to oversee various business units. The intent was clear: inject new expertise into operations and accelerate the company's growth and sophistication.

However, as the ancient proverb warns, 'The road to hell is paved with good intentions.' This decision would soon prove to be a double-edged sword, slicing through the very fabric of what had made WayCool successful.

Some of the newly hired professionals, eager to make their mark and attract customers and vendors, implemented aggressive discount schemes and marketing programmes. On paper, these tactics drove impressive short-term growth. The numbers looked good, and for a while, it seemed like the company was on an unstoppable upward trajectory.

But beneath the surface, trouble was brewing. These strategies, while driving short-term growth, were not sustainable. They began draining financial resources at an alarming rate. The company that had been built on the principles of efficiency and sustainability now found itself burning through cash like a wildfire through dry grass.

Karthik, who had always prided himself on his hands-on approach and deep understanding of the business, found himself increasingly disconnected from the day-to-day operations. The situation came to a head in April 2022.

As Karthik sat through the annual business planning exercise, the rosy projections presented by his team seemed disconnected from the reality he was beginning to perceive. The numbers didn't add up. It was at this moment that he realized WayCool needed to change course – and fast.

He began pushing his team towards profitability, but it was like trying to turn a massive ship with a small rudder. Despite his efforts, the company struggled to gain traction on its profitability goals. By October 2022, the consequences of the aggressive expansion and marketing strategies had become painfully apparent. WayCool, once a beacon of innovation and sustainability in the agri-tech space, now found itself in financial straits.

It was time for Karthik to step back into the trenches. The entrepreneur who had started his journey disassembling toy cars now had to disassemble and rebuild his own company. He took decisive action, ready to make the tough decisions necessary to steer the company back towards profitability.

The period that followed was one of intense introspection and painful but necessary decisions. WayCool began a strategic retreat, exiting several areas it had expanded into, particularly those outside South India. Karthik and his team had to confront an uncomfortable truth – they didn't fully understand how markets outside the South operated. The dream of becoming a pan-India player had led to financial losses, and now they had to make the difficult decision to wind down their presence in states like Maharashtra.

But the pruning didn't stop there. The company also withdrew from segments that were absorbing a lot of working capital and were difficult to control from their Chennai headquarters. The long-distance trade of vegetables and fruits, which had seemed like a natural extension of their business, proved to be a logistical nightmare.

In a significant shift, the company moved away from its direct distribution model to one that utilized Carrying and Forwarding Agents (CFAs) and distributors. This change allowed it to drastically reduce the number of warehouses from a peak of 55 to just 11. Five of these warehouses were used as hubs, with operations also running directly from mills to distribution points. The new model effectively created a network of about 110 distributors who functioned as extension warehouses but operated independently.

Even in existing warehouses, such as those in Vijayawada and Hyderabad, WayCool moved to a CFA model, where a third party runs the warehouse. This strategy allowed it to variabilize its costs, leveraging established processes while outsourcing the operational aspects. It was a return to the lean, efficient model that had made WayCool successful in the first place.

As part of this restructuring, Karthik made another bold move. The technology team, which had been instrumental in building the company's cutting-edge inventory management and supply chain systems, was spun off into a separate company called Sensa. This new entity began servicing external clients, generating additional revenue and helping cover the costs of operating the tech team. The logic was to offer their technology solutions to clients outside South India who had similar needs, thereby expanding the company's reach and impact beyond its core business.

These strategic changes, while difficult to implement, began to bear fruit. By mid-2024, at least three divisions of the company had turned profitable, with the remaining divisions expected to follow suit by October or November. The objective shifted from merely achieving EBITDA (earnings before interest, taxes, depreciation, and amortization) profitability to becoming cash profitable – a significant milestone for a company that had been burning through cash just a couple of years earlier.

Through this tumultuous period, Karthik's perspective on fundraising and growth underwent a profound transformation. The man who had once chased aggressive expansion now began to question the need for constant capital raising. After nine years of operation, he believed it was time for WayCool to grow through self-generated cash rather than continual dependence on external funding. It was a return to basics, a reminder that sustainable growth comes from within.

As he emerged from this period of turmoil, Karthik found himself reevaluating many of the assumptions that had guided the company's growth. The conventional wisdom about market dynamics in India – the perception of big cities as trendsetters and the dominance of national or international brands – was being challenged, particularly in southern states like Tamil Nadu.

To his surprise, Karthik noticed that regional brands held sway in these markets. They understood local customers exceptionally well, a realization that helped explain WayCool's initial success in its first few years, particularly in tier 2 and tier 3 cities of southern India. Contrary to what many might expect, he discovered significant wealth in these smaller cities, with consumers willing to pay a premium for authentic, high-quality products.

This insight led to a renewed focus on WayCool's roots as a regional brand. Karthik recognized that as the company had expanded and brought in outside professionals, it had begun behaving more like a national brand, losing touch with its core customers. The need to return to first principles and learn from successful regional brands became apparent.

Amidst these changes, Karthik also noticed an emerging trend in tier 2 and tier 3 cities: the rise of local hypermarkets. Unlike national chains, these are locally owned and operated ventures, often set up by influential local entrepreneurs. These hypermarkets were becoming weekend entertainment destinations for families in smaller towns, offering not just shopping but also recreational spaces and food courts.

This trend, quietly spreading across South India, presented a new opportunity. The company found that its brands were readily accepted in these hypermarkets, offering a path to expand and saturate the market while leveraging its understanding of local sensitivities. It was a reminder that sometimes the best opportunities lie not in far-flung areas but right in your backyard.

With WayCool navigating these challenges and opportunities, Karthik reflected on his entrepreneurial journey. The past few years had been a crucible, testing his resolve and forcing him to confront hard truths about his leadership and the company he had built. But from this crucible, a stronger, more focused organization was emerging – one that was true to its roots yet poised for sustainable growth.

As the dust settled on the company's restructuring process, Karthik Jayaraman found himself with a wealth of hard-earned wisdom. The journey from near catastrophe to renewed focus had been gruelling, but it had also been illuminating. Like a farmer after a particularly challenging season, he now had insights that could only be gained from facing adversity.

One of the most profound realizations was the danger of delegating without oversight. Karthik had learned the hard way that bringing in outside professionals, while necessary for growth, came with its own set of risks. Many of these new hires, he observed, tended to optimize for their own career metrics rather than the company's overall well-being. It was a sobering lesson in the importance of maintaining a connection to the company's core operations, no matter how large it grew.

This lesson led Karthik to make some unconventional decisions. In a move that raised eyebrows in corporate circles, he began bringing back employees who had been sidelined by the new professionals, putting them back in charge of key business areas. These were the people who understood the company's mission at a fundamental level, who had been there during the lean years and shared the company's values.

Through these experiences, Karthik came to value a particular type of team member: those who were willing to challenge him and engage in collaborative problem-solving. He recognized that these were the people who stuck by the company through thick and thin, embodying the true spirit of WayCool's mission. It was a reminder that in the world of startups, loyalty and shared vision often trump impressive resumes.

In the backdrop of this period of turbulence, Karthik's management philosophy underwent a significant transformation. He observed that the company's brand had established itself in the market, creating a pull that hadn't existed before. This shift was evident in the changing behaviour of distributors, who were now more willing to pay cash for products and work on smaller margins per kilogram rather than demanding high percentage-based margins and credit terms.

This evolution allowed them to focus on fewer, more impactful initiatives without compromising on unit economics or negotiation strategies. Karthik noticed that the distributors and millers who had engaged in tough negotiations with them were the ones who stayed and continued to operate smoothly. Those who had supported them with extended credit during tough times were now reaping the benefits as the company emerged from its challenges. It was a testament to the importance of building strong, mutually beneficial relationships in business.

As WayCool moved forward, Karthik implemented several changes to refocus the company on its core strengths and values. One significant shift was moving away from a headquarters-centric model to a field-centric one. He observed that the company had become too focused on its corporate offices, leading to a disconnect from the realities of field operations.

To combat this trend, Karthik made the bold decision to close the company's last remaining office and move all operations to the warehouses. He also decided to end the work-from-home policy, emphasizing the importance of being in the field for a business that deals with people from the farm fields. It was a return to the company's roots, a reminder that in the agricultural sector, success is often found in the soil, not in air-conditioned offices.

As for WayCool's future direction, Karthik is committed to the company's core strategy of focusing on South India and building strong regional brands. He sees the company's role as creating value through efficient supply chain operations and capturing that value through strong branding. It is a vision that combines technological innovation with a deep understanding of local markets.

Karthik identified potential for growth in value-added products, particularly in categories like rice, lentils, spices, and dairy products and noted the trend towards ready-to-cook, convenience-based foods in markets with similar GDP levels to South India, such as Malaysia or Thailand. However, he recognized the need to develop a unique portfolio of such products tailored to Indian tastes and preferences. It was an opportunity to blend tradition with innovation, creating products that resonated with the changing lifestyles of Indian consumers.

Drawing from his experiences, Karthik shared a set of guiding principles for entrepreneurs entering the agri-space. First and foremost, he emphasized the importance of proving a business model works at a small scale before attempting to expand, even if the initial profits are modest.

He stressed the danger of compromising on unit economics while scaling, noting the difficulty of recouping investments if a company loses its price point and starts operating with negative unit economics. 'In the race to capture market share, don't leave your profits at the starting line' has become one of his favourite pieces of advice to young entrepreneurs.

Karthik also underlined the importance of finding the right market at the right price point, rather than trying to artificially create demand through low prices. He pointed out that in a country as vast and diverse as India, there is always a market for products at the right price point. The key, he believes, is in identifying and serving that market effectively.

One of the crucial lessons Karthik has learned is the danger of applying a one-size-fits-all approach to scaling in the agricultural sector. He recognized that while certain principles might be universal, a specific model needs to be tailored to each geography. It was the stark differences between states like Tamil Nadu, Andhra Pradesh, Kerala, and Karnataka that underscored for him the need for a nuanced, localized approach to expansion.

This led Karthik to question the pursuit of becoming a national brand. He observed that in India, especially for ethnic products, regional brands often dominate and are likely to continue doing so for the foreseeable future. He believes that national brands in this space might eventually evolve into owners of multiple regional brands rather than unified national entities. It's a perspective that challenges conventional wisdom but resonates with WayCool's experiences.

Karthik's thoughts on fundraising and capital allocation also evolved during this period. He began to advocate for a more measured approach to seeking external capital, encouraging entrepreneurs to carefully consider whether they truly need outside investment to grow their businesses, since the pressure to scale rapidly after taking on external capital could lead to compromises that ultimately reduce both profitability and personal freedom for the entrepreneur.

Instead, he considers bootstrapping as a preferable approach, suggesting that entrepreneurs tap into government schemes from institutions like the National Bank for Agriculture and Rural Development (NABARD), even if the process is slower and more complex. For those who do seek external funding, he recommends waiting until the company has achieved significant revenue and stability before considering private equity investments. 'Build a house with a strong foundation before you start adding floors,' he would always advise.

Karthik has learned that growth in the agri-sector would not always follow the smooth, upward trajectory that investors often desire. He is aware that there will be times when the company might need to hold back supply or refuse compromises pushed by customers, leading to choppy growth patterns, a realistic view of an industry subject to the vagaries of nature and market fluctuations.

Reflecting on the most challenging aspects of his entrepreneurial journey, Karthik spoke candidly about the emotional toll of watching the company he built struggle under the management of those he had trusted to improve it. He acknowledged the difficult reality that emerging from such situations often required further dismantling of existing structures before rebuilding could begin. It was a reminder of the personal cost of entrepreneurship – the sleepless nights and difficult decisions that often go unseen.

Despite the tumultuous journey, Karthik remained committed to WayCool's mission. He found motivation in the loyal team members who continued to believe in and build the company, recognizing that their dedication necessitated his own continued involvement and leadership. It was a testament to the power of a shared vision and the importance of building a team that believes in the company's mission.

As WayCool continues to evolve and adapt, Karthik's experiences offer valuable lessons for entrepreneurs in the agri-tech sector and beyond. His journey underscores the importance of staying true to one's vision while remaining flexible enough to adapt to changing market conditions. It highlights the delicate balance between growth and sustainability, as well as the critical role of understanding local markets and consumer behaviour.

Looking to the future, the company is poised to continue its mission of revolutionizing India's agricultural supply chain. With its renewed focus on regional strengths, value-added products, and technological innovation, the company is well-positioned to navigate the challenges and opportunities that lie ahead in India's dynamic and complex agri-food sector.

# 4

# The Accidental Aquaman: Rajamanohar, Aquaconnect

*I*N 2012, THE GRAND *hall in Davos, Switzerland, was filled with global visionaries, world leaders, diplomats, and renowned thinkers from around the world, all converging for the World Economic Forum. The air buzzed with discussions about the future of humanity, strategies to tackle pressing global challenges, and visions to shape a better tomorrow. Among this powerful gathering was Rajamanohar, a thirty-one-year-old entrepreneur recognized as one of the 'Young Global Leaders' for his leadership and contributions in the field of mobile communication. His innovative efforts had brought much-needed connectivity to millions who had once been disconnected from the wider digital world. By delivering digital services through mobile phones, Raj had helped communities embrace the power of technology. For him, this moment of recognition was not just about celebrating past achievements but a call to push further – creating more accessible, impactful tech solutions that could reach even more people.*

*As Raj stood amidst the global elite in Davos, he could never have imagined the unexpected turn his entrepreneurial journey was about to take. The path from mobile technology pioneer to revolutionizing the centuries-old seafood industry was not one he had envisioned. Yet, the skills that had brought him to this prestigious gathering – his insatiable curiosity, knack for problem-solving, and ability to leverage technology for social impact – were about to find an entirely new application.*

# The Accidental Aquaman

*In the world of entrepreneurship, it's often the unexpected encounters that set us on paths we never imagined. For Rajamanohar Somasundaram, known to many as Raj, a serendipitous meeting several years after Davos would transform him from a successful mobile technology entrepreneur into a pioneer in the centuries-old seafood industry. This is the story of how Raj became the 'Aquaman' of India's aquaculture sector, founding Aquaconnect to bring cutting-edge technology to the age-old practice of fish farming.*

Born in the small town of Chidambaram, Tamil Nadu, Raj stood out from an early age for his exceptional academic performance and insatiable curiosity. More than just a stellar student, he thrived in the spotlight, excelling in public speaking and various extracurricular activities.

As he approached the end of his school years, Raj faced a pivotal decision. His father envisioned a future in medicine for him - stable and respected. However, the young man felt an undeniable pull towards creativity and innovation. His mind wandered to architecture, a field where he could merge artistry with functionality, transforming imagination into reality. Despite initial hesitation from his parents, he followed his heart and chose to study architecture.

This decision would prove crucial in shaping Raj's approach to business. His architectural education instilled in him a balance between engineering principles and user-centric design, teaching him to create environments that seamlessly facilitated the needs of all stakeholders. Perhaps most crucially, he developed a deep appreciation for empathy in design, learning the importance of understanding users' needs through research and observation.

Raj's thirst for knowledge led him to pursue postgraduate studies at the prestigious IIT Kanpur. Upon graduating in 2004, he joined a major information technology company, becoming part of an internet banking solutions project. This was a time of rapid technological change in India.

The internet was still in its infancy, most people relied on dial-up connections, and Nokia phones dominated the market, with touchscreen devices still years away from widespread adoption. He found himself at the forefront of this digital revolution, gaining valuable insights into the transformative power of technology.

It was during this time that a chance encounter would plant the seeds for his future entrepreneurial journey. Working late nights at the office, Raj often found himself chatting with the security guards during coffee breaks. One evening, during their usual conversation, a guard asked, 'So, what do you actually do?' Raj smiled and explained how he and his team were putting banking on the internet and mobile phones.

The guard nodded curiously and then quickly exclaimed, 'You people are always building things that people in America and other countries get to use. But what about us, right here? Why aren't Indian software engineers building something for India?'

Raj paused, taken aback by the honesty of the question. It wasn't a criticism; it was a cry for inclusion. At that point in time, India had less than 3 per cent internet penetration, but mobile users made up about 10 per cent and were exponentially growing. He realized the glaring gap between the technologies he built and the people they could truly impact.

This conversation sparked a transformation in his thinking. As he continued his work in the information technology sector, his mind buzzed with possibilities. He envisioned mobile phones not just as communication devices but as gateways to services and opportunities for millions of Indians who might never own a computer.

While grappling with these ideas, Raj reconnected with his former IIT Kanpur classmate, Vinay. Their discussions would set the wheels of entrepreneurship in motion. Vinay, who had been exploring business opportunities, shared his vision of creating mobile-based entertainment. This concept resonated strongly with Raj's desire to leverage mobile technology for wider impact.

Their discussions quickly focused on the potential of mobile gaming. While touchscreen phones and app stores were still years away, simple games like Snake were already captivating millions of mobile users. However, the duo saw an opportunity to go beyond mere entertainment. They both envisioned educational games that could engage users and provide valuable learning experiences – an emerging concept called 'edutainment'.

This idea aligned perfectly with Raj's vision of using mobile technology to impact lives positively. Both believed that by developing engaging educational games and selling them for an accessible price of ₹40 per user, they could not only tap into the growing demand for mobile entertainment but also contribute to education and empowerment.

This vision culminated in the inception of their first company together, Hexolabs, in 2005. Their aim was clear: to develop 'edutainment' games that would capitalize on the burgeoning mobile gaming trend while also serving the higher purpose of education and empowerment. What set Hexolabs apart was the founders' deep understanding of their target users. They adopted a user-centric approach, spending considerable time observing and interacting with potential customers. This methodology, which would become a hallmark of Raj's entrepreneurial journey, gave them valuable insights into user needs and preferences.

However, the mobile applications (popularly known as apps) landscape of 2005 presented formidable challenges. In the era before app stores and smartphones, it was a Herculean task to develop mobile apps and games. The team had to hard-code games directly into phone operating systems, learning coding from scratch. Despite these obstacles, they persevered, developing ten meticulously crafted and rigorously tested games.

As they neared the completion of their games, the founders faced an unexpected hurdle: distribution. In 2005, games came preloaded on branded feature phones, and users couldn't download or install new ones. So the team approached leading global phone manufacturers with a visionary proposal to create a common ecosystem for apps and games – a concept that eerily foreshadowed today's app stores.

But the pioneer mobile handset manufacturers weren't ready for such a paradigm shift, focusing instead on hardware tweaks – experimenting with button shapes, skin colours, and product forms. Hexolabs was ahead of the curve, a position that often comes with its own set of challenges. Their inability to secure a distribution channel ultimately spelled the end of their ambitious project. This setback imparted a valuable lesson: groundbreaking innovation alone isn't enough; the entire ecosystem must be primed and ready to embrace change.

While Hexolabs struggled, the Indian telecom sector was undergoing a revolution. New telecom players entered the market, sparking intense competition and making India one of the world's most affordable mobile phone markets. This transformation fascinated Raj, who began studying mobile phone adoption patterns and how service providers could entice users.

He observed how mobile operators were adopting strategies from the Indian FMCG sector to drive mass adoption. Much like how Velvette shampoo had transformed the market by offering 10 ml sachets for just ₹1, making the product accessible to millions, telecom companies were packaging their services into small, affordable bundles to attract the average Indian consumer. Faced with the challenge of operating in a market with the lowest ARPU (Average Revenue Per User) in the world, Indian operators turned to value-added services as a lifeline to boost profitability.

Raj coined the term 'ABC package' – a bundle of astrology, Bollywood, and cricket services – that became instrumental in driving content consumption among everyday mobile users. These cultural touchpoints tapped into the collective psyche of Indian masses, sparking a wave of mobile engagement that was both affordable and relatable.

Inspired by these observations, he envisioned an SMS-based search engine, 'SMS Wiki', allowing users to access information without internet access. This idea aimed to offer affordable mobile phone value-added services for bottom-of-the-pyramid users, further democratizing access to information.

While developing this concept, the entrepreneurial spirit pushed him to explore opportunities beyond India. Raj identified Africa as a market with immense untapped potential, particularly in the mobile services sector. In a bold move, he bought a one-way ticket to Nairobi for a market study. For over a month, he immersed himself in the local entrepreneurial ecosystem, interacting with startups, exploring the telecom sector, and studying user behaviour.

Armed with this knowledge, Hexolabs successfully launched its services in African markets, significantly accelerating its growth. Through these experiences, Raj challenged and ultimately debunked the myth that building products for bottom-of-the-pyramid users couldn't be sustainable. His strategic approach to pricing and partnerships with mobile operators proved crucial. Under Raj's leadership, Hexolabs evolved into a fully bootstrapped, highly profitable business, demonstrating the viability of serving underserved markets with innovative mobile solutions.

However, the tech world was on the cusp of a revolution that would dramatically reshape the mobile landscape. The iPhone's launch in 2007 marked the beginning of the smartphone era, rapidly rendering old technologies obsolete. The rise of 3G, app stores, and mobile internet shifted power from operators to users and app developers. Apple's App Store and Google's Play Store further diminished the operators' role in content distribution, directly impacting Hexolabs's business model.

With his keen eye for trends, Raj recognized that the era of operator-driven value-added services was coming to an end. Rather than seeing this as a threat, he viewed it as an opportunity to evolve. He began envisioning a future where mobile technology could empower businesses across various sectors, bridging the digital divide and bringing innovation to traditional industries.

As Hexolabs's core business model faced increasing challenges, Raj again found himself at a crossroads. The lessons learned from his venture, the insights gained from his African expedition, and his deep understanding of user needs were all converging, pushing him to explore new horizons. He realized that his true strength lay in technology, and he aspired to become a tech enabler for brick-and-mortar or real-life businesses.

---

It was 2015, and Raj found himself at a crossroads in the ever-evolving tech landscape. Little did he know that a routine train ride to his hometown would set him on an unexpected path, leading him into an industry he had never considered before.

Seated beside Raj was a man in his early forties, engrossed in a phone call about shrimp sales and price negotiations. Intrigued, Raj struck up a conversation with his fellow passenger once the call ended. What he learned was eye-opening: this man had invested his life savings – a staggering ₹50 lakh – into a shrimp farm.

As the train chugged along, the shrimp farmer shared the myriad challenges he faced daily. From growth issues and poor access to quality farm inputs to difficulties in selling produce at fair prices, the 120-day journey from stocking to harvest was fraught with uncertainties. Raj was captivated. He had always considered aquaculture a small-scale industry, picturing neighbourhood ponds supplying local markets. The reality, he was about to discover, was far more expansive and complex.

This chance encounter ignited a spark in Raj. With the same determination that had once driven him to buy a one-way ticket to Nairobi, he decided to dive deep into the world of aquaculture. For months, he traversed thousands of kilometres along the broken roads of rural and coastal regions in Tamil Nadu and Andhra Pradesh. Under the scorching sun, he immersed himself in understanding the industry dynamics, speaking with stakeholders from small-scale farmers to industry veterans.

Raj's initial research revealed numbers that were nothing short of staggering. To his astonishment, he learned that India was one of the world's largest exporters of frozen shrimp and seafood, with annual exports nearing $5 billion in 2016. The global seafood industry was a colossal $160 billion market, with aquaculture's contribution steadily rising as marine sources declined due to overfishing.

The potential seemed enormous. Globally, one out of every two fish consumed came from aquaculture. In India, the ratio was even higher – two out of three. By 2024, the country was projected to earn around $8 billion from seafood exports alone, with the industry showing nearly double-digit growth every year, outpacing the country's GDP.

As Raj dug deeper, he discovered that seafood played a crucial role in global food security, accounting for about 20 per cent of animal protein demand worldwide. It wasn't just about feeding people; it was a source of livelihood for millions in rural and coastal communities across the world. Moreover, aquaculture stood out as one of the most efficient and sustainable protein generators compared to other sources.

These impressive figures and the clear growth trajectory of the industry convinced Raj that aquaculture was worth a deeper look. The market potential was undeniable, and his entrepreneurial instincts were on high alert. With his interest piqued, Raj knew he needed a more comprehensive understanding of the aquaculture ecosystem.

He immersed himself in learning every facet of the industry, from the basics of fish farming to the intricacies of the global seafood market. His research revealed a complex world, far more nuanced than he had initially imagined.

Raj discovered that aquaculture involved the rearing of aquatic creatures in ponds or cages, beginning with farmers purchasing baby shrimp or fish (known as seed or fingerlings) and stocking them in these ponds. The success of aquaculture hinged on maintaining critical parameters: ensuring proper water quality, implementing effective feeding practices, and managing growth. Proper water quality, with the right levels of pH, salinity, dissolved oxygen, and minerals, was crucial for the stock's growth. Feeding needed to be carefully regulated – overfeeding could lead to pollution in the pond, while underfeeding could result in stunted growth.

But aquaculture, Raj learned, consisted of more than just the production phase. It was a complex value chain that extended to both pre- and post-production stages:

1. **Pre-production**: This involved hatcheries producing baby shrimp or fish, feed manufacturing companies supplying necessary nutrition, and healthcare providers

offering probiotics, feed supplements, and disease control solutions.
2. **Production**: The actual farming process, where farmers raised the aquatic creatures.
3. **Post-production**: After harvest, the aquatic produce needed to be processed, packaged, and transported to markets. This involved cleaning, de-scaling, cutting, and storing the fish or shrimp under the right conditions to maintain quality.

Supporting all of this were various services such as technical assistance, training, and research and development.

However, beneath the veneer of impressive market potential and complex processes, Raj uncovered a web of challenges. Despite being the world's second-largest player in aquaculture, India's industry was hindered by a lack of transparency and efficiency in the value chain. At the heart of this complex ecosystem stood the farmer, facing the most acute pain points:

1. **Production Challenges**: Farmers grappled with daily issues in maintaining optimal conditions for their aquatic stock.
2. **Knowledge Gap**: Many struggled to access information on better farming practices.
3. **Input Acquisition**: Obtaining quality seeds, feed, and other necessary inputs was often difficult.
4. **Market Access**: Farmers faced significant hurdles in selling their harvest at fair prices.
5. **Financial Uncertainty**: Fundamental questions remained unanswered for many farmers: Who should they sell to? Where should they sell? What's the right price for their harvest?

Beyond the farmers, other players in the value chain faced their own set of challenges:

1. **Working Capital**: Many stakeholders struggled with cash flow issues.
2. **Traceability**: The lack of a transparent system made it difficult to track the journey of seafood from farm to plate.
3. **Sustainability**: Environmental concerns and the need for sustainable practices added another layer of complexity.

As Raj delved deeper into these issues, he realized that while other sectors had undergone digital transformations, aquaculture had been left behind, still grappling with age-old inefficiencies. It was this stark contrast between the industry's potential and its current state that led to a moment of clarity for Raj. With his expertise in building tech-driven solutions, he visualized how technology could be a game changer for the sector if leveraged effectively.

With this comprehensive understanding, Raj recognized that focusing on farmers would be the most effective starting point for his venture. The growing nature of the industry was intriguing, and the glaring gaps in the value chain gave him the confidence to step in. Here was an opportunity to make a real difference, applying his tech expertise to an industry ripe for innovation. In 2017, he incorporated Aquaconnect, with a mission to bring transparency, predictability, and efficiency to the whole aquaculture value chain, enabling more effective linkages between stakeholders.

Eschewing the typical tech startup approach of immediately developing an app, Raj opted for a grassroots strategy. He deployed a team of aquaculture experts as field officers in the aquaculture hubs of Chidambaram, Sirkazhi, and Nagapattinam in Tamil Nadu. These officers became Aquaconnect's eyes and ears on the ground, working closely with farmers to understand their problems and provide practical solutions. They offered guidance on best practices, helped monitor pond conditions, and advised on feed management.

To complement the field team's efforts, Raj established a call centre. This hub served as a central point for gathering information and identifying problems more efficiently. The synergy between the on-ground officers and the call centre team created a comprehensive support system for farmers, addressing their issues promptly and effectively. This hands-on approach allowed Aquaconnect to gain invaluable insights into the farmers' daily struggles and needs.

As Aquaconnect's understanding of the aquaculture ecosystem deepened, two critical areas emerged as prime candidates for improvement: farm input solutions and post-harvest linkages. Recognizing the potential of technology to address these challenges, Raj and his team set to work on solutions.

To gain a global perspective on aquaculture practices, Raj travelled to Norway in 2018. For several weeks, he immersed himself in the Norwegian aquaculture industry, interacting with experts and observing local practices. The advanced technology and infrastructure he witnessed fascinated him, showing him first-hand how automation and sustainable practices could revolutionize the industry.

Armed with these insights, Aquaconnect launched an online marketplace the same year, where farmers could directly order farm inputs. However, as a pioneer startup, they faced challenges in acquiring farm inputs directly from manufacturers. Undeterred, Raj forged partnerships with distributors to ensure a steady supply of farm inputs through their online platform.

The post-harvest phase presented its own set of challenges. Raj observed a glaring lack of transparency that affected not only the farmers but also seafood buyers. The existing system was dominated by multiple layers of middlemen, with neither the farmer nor the buyer directly involved in transactions. This opaque process often resulted in unfair pricing and quality issues.

To address this, Aquaconnect began onboarding seafood processors into their network. This strategic move was designed to infuse transparency into the post-harvest market and organize it more efficiently. By acting as a sourcing partner on the ground, the company could now assist farmers in selling their harvest more transparently while helping processors source quality seafood in the required quantities.

This dual approach – facilitating input supply and streamlining post-harvest processes – laid a solid foundation for Aquaconnect's business model. The startup now had a workforce that consistently touched base with farmers, helped them purchase inputs, and assisted in selling their harvest through the platform. Buoyed by their success in Tamil Nadu, operations expanded into Andhra Pradesh, India's leading aquaculture state, with a focus on deepening their presence there.

As the company grew, Raj and his team began to envision their next strategic move. Drawing from their accumulated learnings, they conceived the idea of improving production efficiency through a data-driven farm advisory. This led to the development of an innovative app that would revolutionize aquaculture management.

The app, christened 'FarmMOJO', was designed to be a comprehensive farm companion. It allowed farmers to manage all their key activities, from stocking to harvest, capturing crucial data on water quality, daily feeding, growth sampling, and animal health. But it wasn't just a data collection tool; FarmMOJO used this information to provide personalized alerts and suggestions to improve water quality, optimize feeding patterns, and maintain animal health.

The app tracked growth parameters, provided daily market prices, offered information on the availability of farm inputs, and even included feed calculators. In essence, FarmMOJO was designed to be a digital assistant, working alongside farmers throughout the aquaculture cycle and helping them make informed decisions to optimize input usage and improve productivity.

Raj envisioned FarmMOJO as 'a super app for seafood', an integrated digital ecosystem for all stakeholders in the value chain. Farmers could use it to manage their farms, order inputs, and sell their harvest. Farm input manufacturers could gain real-world feedback on their products, creating a valuable feedback loop. Seafood buyers could plan their procurement more effectively and reach out directly to farmers.

When the app was piloted, Raj was ecstatic with the overwhelmingly positive feedback from the farmers. Encouraged by this response, the team officially launched FarmMOJO in mid-2019. It was a revolutionary concept that promised to transform the entire aquaculture industry.

However, the journey of innovation is rarely smooth, and FarmMOJO faced its share of challenges. Despite good initial traction in terms of installations, the team was disappointed to discover low adoption rates among the farmers. The number of daily active users was below expectations. Determined to understand the root cause, Raj and his team went back to the field.

Their conversations with farmers revealed a complex set of issues. Many felt overwhelmed by the effort required to use the app. They found the learning curve steep and were concerned about data privacy. Some simply didn't see the immediate value in entering data into the app, preferring their traditional methods of record-keeping. The team realized they had underestimated the farmers' attachment to familiar routines and overestimated their readiness to embrace new technology.

This setback was a valuable learning experience for Raj and his team. They went back to the drawing board, tweaking FarmMOJO to reduce the inputs required from farmers and make the app more intuitive. However, the real breakthrough came from an unexpected source – the COVID-19 pandemic.

The onset of lockdowns in 2020 dramatically changed the landscape. With movement restrictions in place, farmers had no choice but to turn to digital solutions, resulting in a significant increase in FarmMOJO usage. Recognizing the unique challenges posed by the pandemic, Aquaconnect quickly adapted its services. It launched a helpline to assist farmers with remote farm advisory, clarify government lockdown regulations, and provide access to markets for selling produce. The company even offered free delivery services to farmers, regardless of order quantity, ensuring essential supplies reached their doorsteps.

These initiatives provided critical support to farmers during the pandemic and showcased the startup's ability to pivot and respond to crisis situations. However, as lockdowns lifted, many farmers reverted to their previous practices, and app usage declined once again. This regression was a blow to Raj and his team, who had hoped the pandemic-driven adoption would lead to lasting change.

Disappointed but determined, they pressed on with their innovations. They integrated FarmMOJO with a chat interface reminiscent of WhatsApp, enabling farmers to upload photos for data conversion. Despite these improvements, adoption rates remained below expectations.

This experience offered Raj a sobering lesson about the pace of technological change in traditional sectors. He recognized that farmers couldn't be expected to transform their long-standing habits overnight. Instead of trying to replace their existing methods entirely, Aquaconnect needed to find ways for its solutions to complement and gradually enhance established farming routines.

While grappling with the challenges of app adoption, the company also faced hurdles in its online marketplace for aquaculture products. Despite its apparent potential, the business proved less lucrative than initially anticipated. The high-volume, low-cost nature of many aquaculture products, combined with the logistical complexities of shipping to coastal and interior parts of the country, often resulted in logistics expenses exceeding profit margins.

As a funded startup, Aquaconnect had the option to continue these online operations despite the challenges. However, this approach conflicted with one of Raj's core principles: 'Don't grow at all costs.' He had instilled a culture of prioritizing healthy business transactions for both the customer and the company, emphasizing the importance of maintaining positive unit economics. This principle was deeply embedded at all levels of the organization.

Faced with these challenges, Raj and his team returned to the drawing board, determined to find a solution that would amplify Aquaconnect's reach while addressing the core issues in the aquaculture supply chain. After months of intensive research and brainstorming, they arrived at a pivotal insight: the key to transformation lay not directly with the farmers, but with the farm input retailers.

These local entrepreneurs, deeply embedded in their communities, speaking the same language and following similar customs, were already trusted advisers to farmers on everything from feed selection to best farming practices. Raj envisioned transforming these retailers into 'anchoring partners' for Aquaconnect, enabling last-mile delivery of the company's services in every village.

The advantages of this approach were manifold. First, it solved the scalability issue that had plagued their previous model, enabling the organization to expand its impact across diverse geographies with greater efficiency. Second, retailers were constant touch points for farmers, sharing their experiences with different products and practices and helping farmers make informed decisions. It was a perfect community engagement-driven model as the advisory was directly given to the farmers through the retailer. Lastly, working with retailers simplified logistics, as the company could focus on delivering products to individual aqua partners who in turn could deliver services to around a hundred network farmers.

As Raj and his team immersed themselves in the world of aquaculture retailers, they uncovered a new set of challenges. These entrepreneurs in rural and coastal communities, often with limited exposure to technology and formal education, were struggling to meet the diverse needs of their farmer clients. They had to stock multiple products from different brands and vendors, navigate relationships with various distributors, and grapple with minimum order quantities that forced them to overstock or incur high logistics costs.

Moreover, many retailers lacked access to a wide range of brands, leaving them at a disadvantage when farmers requested specific products. The need to order in large quantities also led to problems with storage, wastage, and tied up their capital. It became clear to Raj that these retailers needed an aggregation platform – a one-stop-shop where they could order any input material, from multiple brands, in any quantity, to effectively meet farmers' needs.

In response to these insights, Aquaconnect launched the innovative 'Aqua Partner' programme. This initiative was designed to onboard rural entrepreneurs into their network, transforming them into empowered agents of change in the aquaculture industry by optimizing service delivery. The company forged partnerships and signed MOUs with manufacturers, offering retailers access to a wide array of input brands and quantities with a single click.

They strategically relocated warehouses and restructured key activities to align with their new strategy. Crucially, they eliminated minimum order quantities for Aqua Partners, optimizing logistics by covering multiple retailers along a single delivery route.

The Aqua Partner programme proved to be a resounding success. Within 18 months of its launch, the company had onboarded over 400 retailers across four states – Andhra Pradesh, Odisha, Tamil Nadu, and Gujarat. This rapid expansion validated Raj's vision and demonstrated the hunger for such a solution in the market.

As the network grew, the Aquaconnect team soon identified another critical challenge: access to working capital. Many retailers, despite their crucial role in the supply chain, lacked the financial history or business data that traditional institutions required to assess creditworthiness. This limitation on their ability to secure loans often stunted their business growth, constraining them to operate within the bounds of their available capital.

Never one to shy away from a challenge, Raj saw this as an opportunity to further innovate. Aquaconnect introduced the groundbreaking AquaCRED programme, a 'Buy Now Pay Later' option that enabled retailers to access working capital and repay it within 30 days, interest-free. This initiative involved establishing anchor partnerships with financial institutions to help retailers secure formal finance. A successful pilot project in Andhra Pradesh paved the way for broader implementation.

The mechanics of AquaCRED were elegantly simple yet transformative. Retailers could order materials through

Aquaconnect's platform, which would then send the invoice directly to the partner bank for payment. This system ensured that the funds were utilized strictly for business purposes, alleviating banks' concerns about fair utilization of credit. The availability of this formal finance option attracted even more retailers to the Aquaconnect platform, as it eased their daily working capital needs and provided them with a pathway to scale their businesses.

Concurrent with these financial innovations, the company invested heavily in strengthening its last-mile delivery capabilities. This focus on logistics proved to be a game changer, particularly for retailers in remote rural and coastal areas who had long struggled with unreliable supply chains. In a bold move, Aquaconnect introduced overnight delivery as a pilot project in select areas – a service almost unheard of in the industry. Despite the inherent challenges of implementing such a rapid delivery system in often infrastructure-poor regions, the success of the pilot led to its expansion across several areas.

Leveraging its growing scale, the team was able to negotiate better terms with suppliers, passing on the benefits to retailers in the form of lower prices and improved trade terms. The company's ground officers now split their time between supporting Aqua Partners and assisting farmers directly. They helped retailers with inventory management, marketing campaigns, digital lead generation, and even WhatsApp-based customer outreach, all aimed at increasing the business turnover of the Aqua Partners.

The flourishing retail network caught the attention of manufacturers and biotech innovators, previously distant players in the aquaculture ecosystem. This growing interest signalled a potential transformation of the entire value chain. Two key factors drove this surge of attention: first was the impressive network Aquaconnect had painstakingly built. The platform now boasted a vast array of retailers, offering manufacturers an efficient channel to reach their target market. What was once a fragmented, hard-to-access market had become a streamlined pathway for product distribution.

Second, the platform had evolved into a valuable source of market intelligence. The ground team, deeply embedded in the day-to-day realities of aquaculture, actively monitored product demand and performance. This real-time feedback proved invaluable for manufacturers, offering insights to refine their offerings and meet the genuine needs of farmers.

Recognizing the importance of cutting-edge solutions in aquaculture, Raj saw an opportunity to further innovate. The team began actively identifying and onboarding biotech companies, using their Aqua Partner network to launch these innovative products in the market. This curated approach to product selection not only enhanced Aquaconnect's reputation but also accelerated the pace of innovation in the industry.

For biotech startups, the platform became a fast lane to market. Raj marvelled at how a promising startup from Israel, for instance, could now bring their innovative vaccine for aquaculture diseases to the shelves of Aquaconnect's vast retail network within just a month. What once might have taken months or years of market penetration was now achieved in weeks, thanks to the highly integrated platform they had built.

This symbiotic relationship between the platform, retailers, and manufacturers created a virtuous cycle of improvement and innovation. Manufacturers could focus on their core competency of product development, while Aquaconnect handled the complexities of distribution and sales. The platform's rigorous quality checks before introducing new products saved retailers time and effort in product research, making Aquaconnect an increasingly attractive partner for companies seeking to penetrate new markets while maintaining high-quality standards.

The impact of Aquaconnect's evolving model extended far beyond the immediate stakeholders of manufacturers, retailers, and financial institutions. By leveraging the influence and reach of retailers, the company was able to disseminate valuable knowledge and best practices in aquafarming to a vast network of farmers. The retailers, now well-trained and equipped with knowledge on packages of farming, became trusted advisers to farmers, helping them implement new techniques quickly and improve their production.

Having successfully transformed the farm input side of the aquaculture industry, Raj and his team set their sights on the other end of the value chain. Their goal was ambitious yet clear: to leverage their growing network of Aqua Partners to create seamless linkages between seafood buyers and farmers.

This new initiative expanded the role of retailers significantly. No longer were they just suppliers of inputs; they now became pivotal players in providing technical advisory services to farmers and facilitating the procurement of farm output. This evolution transformed retailers into true enablers, bridging the last mile and forging stronger connections with farmers. The enhanced role created a stickiness in the company's relationship with farmers, offering them an added advantage they had never experienced before.

For seafood buyers, Aquaconnect's platform became a game changer. Acting as a comprehensive sourcing platform for fish and shrimp, it eliminated the need for multiple layers of middlemen that had long plagued the industry. Buyers could now simply inform Aquaconnect about their seafood sourcing needs, quantity, and price, and the sourcing problem was solved efficiently and transparently.

Recognizing that the seafood processing industry is highly capital intensive, Raj and his team took their financial innovations a step further. They extended the AquaCRED programme to seafood buyers, introducing a novel solution to ease their daily working capital needs. Under this system, when a buyer ordered seafood from the platform, the partnering bank paid the company directly, rather than the buyer. This arrangement assured banks that the funds would be used for genuine purchases, increasing their willingness to provide loans to buyers using the Aquaconnect platform.

The impact of this financial innovation was immediate and profound. It quickly became a huge win for Raj and his team. Perhaps most significantly, it enabled same-day payments for farmers once they sold to seafood buyers – a radical improvement over the industry norm of payment terms stretching from a couple of weeks to as long as 50 days.

Through these comprehensive efforts, Aquaconnect was steadily transforming the entire seafood value chain. By addressing the challenges faced by all stakeholders – from farmers to retailers to buyers – they were creating a more efficient, transparent, and equitable system.

Throughout this period of rapid innovation and expansion, Aquaconnect steadily grew its geographical footprint. The company spread its operations across the major aquafarming states in India, including West Bengal, Uttar Pradesh, Bihar, and Assam, bringing its transformative approach to new regions and communities.

One of the most persistent roadblocks in any value chain is the lack of data and data-driven decision-making capability. Aquaconnect tackled this head-on, bringing unprecedented transparency to the value chain. By leveraging technology and innovation, they created a platform that enabled retailers to access formal finance, manufacturers to reach new markets, and farmers to improve both the quality and quantity of their produce.

However, despite their successes, one challenge remained stubbornly persistent: getting farmers to input their farm-level data into the app consistently. Raj, ever the innovator, realized that a new approach was needed – one that was both non-intrusive and scalable. The solution came in the form of AI and geospatial technologies, culminating in the development of the AquaSAT platform.

The concept behind AquaSAT was ingenious in its simplicity. The platform would capture pond images through satellite imagery and use AI to mark pond boundaries. However, the execution proved to be far from simple. Obtaining high-resolution images was a significant challenge, requiring Aquaconnect to develop and run sophisticated AI models to enhance image resolution and accurately identify pond boundaries.

These enhanced images were then used to train AI models to not only identify pond boundaries but also determine whether a pond was used for rearing shrimp or other fish. Once validated, the system could predict a pond's days of culture – the number of days the fish or prawns had been in the pond. This data intelligence became the foundation for building insights to predict farm input demand and harvest supply at multiple levels – from individual ponds all the way up to ponds across entire states and even the country – all without dependency on human inputs.

Aquaconnect's approach of combining 'boots-on-the-ground' intelligence with 'eyes-in-the-sky' satellite imagery provided a comprehensive sector-level crop review. It could determine how many ponds were active or inactive in an area, state, region, or country - a move that can significantly improve predictability in the aquaculture industry.

The AquaSAT platform plays a pivotal role in supporting the company's business verticals by leveraging backend predictive intelligence. Through farm input demand prediction, Aquaconnect gains insights into the input requirements across different geographies, ensuring the timely provision of the right inputs, which helps manage inventory more efficiently and better serve farmers in each region. Similarly, using harvest-supply predictions, the company assists seafood buyers in sourcing more effectively, enabling them to plan procurement region-wise and streamline their operations. This entire process is powered by AquaSAT's backend intelligence, driving efficiency and optimization throughout the value chain.

As the platform continued to evolve, Raj and his team explored additional use cases. One particularly promising area was traceability. As data intelligence was captured along the production value chain, Aquaconnect enabled end-to-end verifiable traceability for international seafood buyers under the 'Story of Shrimp' programme. A pioneering move in the global seafood space to empower international seafood buyers to source with confidence knowing its exact origin and journey.

The potential of AquaSAT extended far beyond commercial applications. Its near real-time capabilities have great potential to assist governments, institutions, and policymakers in understanding effective monitoring, utilizing land resources, zoning production areas, and assessing diseases and production patterns on a large scale.

Always with an eye on the bigger picture, Raj saw an opportunity to leverage AquaSAT to address one of the most pressing issues of our time: climate change. Recognizing the importance of sustainability and efficiency in the food value chain, Aquaconnect began exploring ways to use its platform to drive decarbonization in the seafood sector.

The rationale was compelling. Seafood accounts for approximately 20 per cent of animal protein consumption globally. Making its production environmentally friendly with a lower carbon footprint could contribute to the planet's sustainability on a larger scale.

It is essential to capture data to understand the impact of carbon emissions. Through AquaSAT, Aquaconnect aimed to quantify carbon emissions from the industry, providing a foundation for strategizing countermeasures. By providing the industry with the tools to measure and understand its carbon footprint, the company aims to take concrete steps towards sustainability.

This ambitious move positioned Aquaconnect at the forefront of efforts to help the global seafood sector combat climate change and accelerate 'net zero' efforts. This also aligns closely with international seafood buyers who have ambitions to reduce carbon emissions in their seafood sourcing journeys.

The company's success in the seafood value chain allowed it to expand its operations to include seafood processing and exports. After a brief pilot in seafood exports to understand the challenges faced by international buyers and exporters, Aquaconnect created a global seafood sourcing platform to eliminate multiple layers of handshakes and provide end-to-end assistance for seafood buyers worldwide, ensuring transparent quality control while fulfilling their traceable and sustainable seafood procurement needs. Around 30 buyers from the US, Vietnam, Japan, and China have recognized and collaborated with Aquaconnect to source directly from Indian seafood processors and exporters.

As Aquaconnect deepened its collaborations with farmers and retail partners over the years, it became clear that true innovation was needed at the farm input level to address the practical, day-to-day challenges faced by farmers. Armed with farm-level data intelligence, Raj and his team established their own R&D division under 'Dr. Grow', dedicated to advancing aquaculture biosciences and innovating animal nutrition and healthcare formulations. By setting up this division, the company unlocked new opportunities to innovate and help farmers improve their productivity.

What sets Aquaconnect apart from traditional manufacturers, who typically focus solely on production, is its unique approach to leveraging its vast network of retailers and field officers. This direct access to farmers and daily interactions with retailers provide real-time insights from the field. These insights create an efficient feedback loop completed with data-driven testing, allowing the company to act quickly, make iterative improvements to its product formulations, and address the challenges faced by farmers in diverse geographical regions.

By testing products directly on the ground and capturing data at every stage, Aquaconnect can continuously innovate and evolve its offerings, ensuring they remain effective across varying farming conditions. This dynamic feedback system enables rapid adjustments and bridges the gap between cutting-edge biotechnology and practical applications.

Aquaconnect's journey under Raj's leadership defied many of the industry's prevailing myths. While conventional seafood businesses followed a siloed approach, Aquaconnect embraced the power of integration. By focusing on unlocking value across every segment of the seafood value chain and partnering with multiple stakeholders, the company demonstrated that a holistic strategy was key to driving real, sustainable impact.

Today, Aquaconnect stands as India's largest integrated seafood platform with a presence in 8 Indian states, working with over 850 retailers, more than 30 biotech companies, and over 100 seafood buyers. This remarkable growth is a testament to the effectiveness of Raj's vision and approach.

Raj and his team understood that improving production efficiency and reducing the cost of aquaculture required more than just software, IoT devices, and best practices that only marginally improved efficiency. As Raj often emphasized, agri-tech entrepreneurs must recognize that having technology alone is not enough to address the challenges faced by the industry. They must be present in all parts of the value chain and understand that technology is merely an enabler. The true differentiator lies in understanding the intricate dynamics of the entire ecosystem and addressing pain points across the board.

Perhaps one of the most defining aspects of Aquaconnect's success can be traced back to a simple yet powerful principle that Raj embedded deeply within the company's culture: 'Encourage everyone to experiment fearlessly.' This open-ended approach to innovation empowered the team to explore new ideas, take risks, and test assumptions. Failures were seen not as setbacks but as valuable learning opportunities, while successes became blueprints for future growth. This culture of experimentation injected a sense of adaptability and continuous improvement, driving Aquaconnect to constantly evolve and stay ahead in the seafood industry.

But Raj's mission doesn't end with just improving the seafood value chain. By leveraging data-driven insights, artificial intelligence, and cutting-edge biotechnology, Aquaconnect contributes to a larger, more urgent cause: the creation of a transparent, efficient, and predictable global seafood system.

'The future of seafood production,' Raj often says, 'isn't just about feeding people. It's about doing so in a way that preserves our planet and empowers the communities that rely on it for their livelihoods.' He envisions Aquaconnect expanding beyond India's borders, helping to revolutionize the seafood supply chain on a global scale. As part of this vision, the company is also aiming to launch an IPO within the next two years, amplifying its impact and scaling its mission on an international level.

For Raj, Aquaconnect is more than just a business; it's a movement – a movement to show that even the most traditional sectors can be transformed with the right mix of technology, empathy, and bold thinking. As the company continues to evolve, Raj remains steadfast in his belief that innovation is not just about solving today's problems; it's about anticipating tomorrow's challenges and creating a better, more sustainable seafood system for future generations.

# 5

## The Achrekar of Agri-Tech: Hemendra Mathur, Investor and Mentor

NOT EVERYONE IS FORTUNATE. *Some people are compelled to take the only path available to them and make the best of it for the benefit of others. They have no qualms about performing jobs that other individuals would dread or consider tedious. They are the world's silent revolutionaries. They continue to perform their duties, not awaiting accolades or awards. They are the world's most vital cogs, propelling it forward. These individuals are the Achrekars, responsible for the emergence of Tendulkars. Hemendra Mathur is one of these Achrekars. He was instrumental in the evolution of several Sachins, Sehwags, and Kohlis of the agri-tech industry.*

Hemendra Mathur's journey is one of resilience, adaptability, and passion for agriculture. Born into a middle-class family, he was instilled with the belief that education, specifically an engineering degree, was the key to financial stability. With this goal in mind, he dedicated himself to his studies from a young age, preparing diligently for the Pre-Engineering Test, a competitive exam conducted by various state governments in India for admission to engineering colleges.

Despite his best efforts, he faced his first setback when he secured a rank of 646 out of approximately 30,000 students vying for just 650 spots. With only agricultural engineering available and without the luxury of waiting another year to retake the exam, he found himself at a loss as to what to do. The course had a relatively poor employability rate, with the primary work opportunity being in the public sector.

At this pivotal moment, a counsellor, noticing his hesitation, advised him, 'Your degree should be about building a career, not about jobs. After agricultural engineering, you could pursue an MBA or MTech degree if you don't get a job after college. You can flip the circumstances around to your advantage.' These words instilled a sense of confidence in the young student, leading him to enrol in the agricultural engineering programme at the College of Technology and Engineering in Udaipur (CTAE).

Hemendra's time at CTAE was exceptional. Captivated by the campus and its resources, he delved deep into agri-engineering subjects, particularly soil conservation, irrigation, and farm mechanization. His passion for the field grew exponentially, and his dedication made him the class topper, breaking several academic records.

In his final year, his achievements reached new heights. His percentile in the GATE exam was an impressive 99.67, securing an all-India rank 1 for him in agricultural engineering. He also cleared the CAT, becoming the first from his college to be successful in both entrance tests. However, despite these accomplishments, he didn't receive any job offers from the top agri-companies, which left him disheartened. He could not help but feel that someone needed to make the agribusiness sector more alluring to attract and absorb the best available talent. Little did he know that two decades later, he would contribute to the glamorization of the sector, attracting IIT/IIM graduates to it in droves.

With his job prospects slim, Hemendra found himself torn between National Institute of Industrial Engineering

(NITIE; later rebranded as IIM Mumbai), which aligned with his engineering background, and IIM Ahmedabad, which his family saw as a safer bet for job prospects. On the one hand, he believed that an MBA would render his engineering degree irrelevant, as he would be focusing primarily on subjects such as economics, marketing, finance, organizational behaviour, and strategy, rather than the hard-core engineering stuff that he had spent four years mastering.

On the other hand, the IIM degree was too attractive to ignore. He saw it as a more secure path, one that would open doors to a wider range of career opportunities and provide him with greater financial stability in the long run. After much deliberation, he decided to pursue an MBA at IIMA (MBA used to be called PGDM in those days), but he couldn't shake off the feeling that he was being pulled away from his true calling.

IIM Ahmedabad proved to be a transformative experience. Surrounded by exceptional peers, he acquired a wealth of knowledge and honed his leadership skills. The experience of studying there broadened his horizons, making him open to diverse opportunities upon graduation, even as his heart still yearned for a career in the agri-sector.

After IIM, Hemendra's love for market research led him to ORG-MARG, where he immersed himself in consumer insights for nearly two years. Eager for new challenges, he then transitioned to KSA Technopak, a boutique consultancy specializing in retail and consumer goods. It was here that he truly found his niche in consumer research and supply chain management.

Working with major retailers like TESCO, Carrefour, and food chains like Pizza Hut, Hemendra gained invaluable insights into global and local food supply chains. He witnessed first-hand how consumer-retail interfaces enabled these enterprises to develop compelling value propositions that resonated with their customers.

Reflecting on his experiences, he says, 'As investors, we sometimes overlook the founders' understanding (or the lack of it) of the end consumers. However, that is the foundation for building a great business. Despite many founders coming from a tech background, they must spend time with customers and gain an understanding of them. This customer understanding must remain a top priority.'

Hemendra's time at KSA Technopak also instilled in him a deep appreciation of the value of time. He learned that every minute spent on a client's behalf, for which they are billed, needs to create value for them. 'Time is money,' he explains. 'There's no better way to learn this than in a high-pressure consulting job.' This realization fuelled his determination to maximize efficiency and deliver tangible results.

Despite his success in consulting, a nagging feeling persisted that he wasn't utilizing his agricultural knowledge. The fear that his engineering degree would become a mere footnote in his career gnawed at him, urging him to reconnect with his roots in agriculture.

Serendipitously, an opportunity arose when Rabobank extended an offer through a mutual acquaintance. This moment marked a significant turning point, allowing Hemendra to redirect his focus towards his true passion: agriculture.

With renewed determination, he joined Rabobank, one of the world's largest banks focused on the food and agricultural sector, serving agribusinesses. As part of the bank's advisory and research division, he found himself at the forefront of two major projects with the Ministry of Food Processing Industries.

The first undertaking was to draft the Food Processing Policy of India. He immersed himself in the task, engaging in discussions with large corporations and hundreds of small and medium-sized businesses in the food supply chain.

Through these interactions, he gained a deep understanding of the challenges and opportunities faced by the sector in India. He learned that the existing policy framework was riddled with complexities and inefficiencies that hindered the industry's growth, particularly for small and medium-sized businesses.

Hemendra and his team crafted a comprehensive policy report addressing key issues and proposing initiatives to foster growth. The resulting document, upon its release by the government, was hailed as the bible of food processing policy. It proposed measures to create infrastructure, institutions, and a more conducive regulatory environment for businesses of all sizes, including the rationalization of the tax structure and the establishment of food parks. This work laid the foundation for a more business-friendly ecosystem that would attract investment, foster innovation, and ultimately drive growth for the food processing industry in India.

The second project involved establishing the National Institute of Food Technology, Entrepreneurship and Management (NIFTEM) for the Government of India, a crucial step towards nurturing talent and fostering innovation. Working closely with stakeholders, Hemendra helped design a curriculum that would equip students with the skills needed to succeed in the rapidly evolving food processing landscape.

He also played a vital role in the National Horticulture Mission (NHM) of the Ministry of Agriculture. Recognizing that horticulture, which includes fruits, vegetables, and flowers, had immense potential to boost farmer incomes and contribute to the overall growth of the agricultural sector, he threw himself into the initiative with enthusiasm.

The NHM implemented initiatives to support horticulture farmers, including establishing state-of-the-art nurseries, promoting efficient farming techniques, and developing cold-chain infrastructure to reduce post-harvest losses and improve market access.

Initially sceptical of NHM's impact, Hemendra witnessed over time how its efforts led to a significant increase in horticulture production, surpassing that of staples. Farmers who adopted the recommended practices saw their incomes rise, and the horticulture sector emerged as a key driver of agricultural growth, helping lift millions of farmers out of poverty.

As he witnessed the transformative power of the policies he helped shape, his passion for the agricultural sector deepened. Driven by this new-found enthusiasm, he began exploring opportunities to make a lasting impact in the industry.

His search led him to Yes Bank in 2006, where he joined the organization to raise a private equity fund for food and agribusiness. Confident in his expertise and extensive network, the young consultant threw himself into the role with enthusiasm, leveraging his deep understanding of the agri-sector to identify promising investment opportunities and craft compelling pitches to potential investors.

However, midway through 2007, the global financial markets began to experience tremendous stress, and the ripple effects soon reached India. The arrival of the 2008 financial crisis hit like a hurricane, with several large banks, including Lehman Brothers in the US, declaring bankruptcy. Raising funds in this unstable environment became increasingly difficult, especially for a first-time fund manager.

Despite the best efforts of his team, Hemendra faced one of the greatest failures of his life as he struggled to secure investments. It was a humbling experience that tested his resilience and forced him to re-evaluate his approach. The once-confident consultant found himself grappling with rejection and uncertainty as he navigated the challenging world of fund management.

Undeterred by this setback, he remained committed to finding ways to drive positive change and growth in the agri-sector. After much soul-searching and reflection, Hemendra decided to join SEAF India Investment Advisers, once again taking on the challenge of raising funds for agribusiness. As the global financial crisis subsided, he hoped for a more favourable outcome, but the obstacles continued.

Investors questioned his lack of investment experience, expressing doubts about entrusting him with their money. Some even pointed out that his background as a consultant and banker did not necessarily translate to being a successful investor. He found himself at a loss, struggling to understand why his years of expertise and deep understanding of the agri-sector were not enough to win over potential investors.

It was during this trying time that Hemendra received invaluable advice from an investor. 'Hemendra,' the adviser said, 'it is natural for people to see you as a consultant or banker but not as an investor. You will never be able to justify yourself as an investor to those people. Do one thing. Imagine that you have money today. Create a list of ten companies that you think we should invest in and explain why we should invest in them.' This simple yet profound suggestion sparked a light-bulb moment for Hemendra and his team.

With renewed focus, he met with approximately 30 agribusiness founders in his network, meticulously gathering information about their firms, competitive advantages, financial data, revenues, and business strategies. Narrowing the list down to 10 promising candidates, he crafted a compelling portfolio. Armed with this new approach, he pitched the portfolio to investors, confidently stating, 'If I had money, I would probably invest in these companies,' and proceeded to explain his rationale.

To his surprise and delight, the investors were captivated by his pitch. They appreciated the depth of his research and the clarity of his vision. Over the next four years, Hemendra along with his team raised about ₹250 crore in funds and deployed the funds in nine deals across the food supply chain. His investments spanned a diverse range of businesses, including food ingredients, animal feed, seeds, fertilizer, spices, milling, restaurants, and organic foods.

As he navigated the evolving landscape of the agri-industry, Hemendra noticed a significant shift beginning in 2016. A new generation of technology-savvy entrepreneurs began seeking his guidance, and agri-tech startups started to proliferate. These startups were markedly different from traditional family-owned agricultural enterprises. Recognizing the potential of this emerging trend, he embraced his role as a mentor and evangelist, sharing his insights with the startup founders of this new era.

What started as an extracurricular activity quickly transformed into a passion. He immersed himself in their world, listening to their ideas, offering them advice, and accompanying them to the fields to assist with experiments and prototypes. Witnessing first-hand the transformative power of technology in agriculture, he realized that its widespread adoption would soon revolutionize the sector, presenting unprecedented growth opportunities.

Inspired by this realization, Hemendra set his sights on investing exclusively in agri-tech ventures. However, he discovered that the SEAF fund, with its focus on small and medium enterprises, didn't align with his vision of supporting startups. Undeterred, he explored other avenues, only to be shocked by the lack of funds willing to invest in agri-startups. The prevailing perception was that these ventures were too risky, and the startups, being relatively new, had yet to establish a track record of success.

Faced with this predicament, Hemendra found himself in a dilemma. The only path forward seemed to be starting on his own, a daunting prospect at 44, with a family to support and the uncertainties of the early months without steady cash flows. Yet, his entrepreneurial spirit refused to be silenced. He knew that if he didn't take the leap now, he might never have another chance.

After much soul-searching and numerous discussions with his family, he made the most difficult decision of his life. He left the comfort and stability of a full-time job, venturing into the unknown with nothing but his name and reputation to back him. It was a defining moment, a testament to his unwavering commitment to the agri-industry and his belief in the potential of startups to drive change within it.

As he immersed himself in the world of agri-tech entrepreneurs, Hemendra witnessed the challenges they faced in validating ideas and prototyping innovations. He realized that the process of seeking feedback from multiple stakeholders often led to delays in product launches, and in some cases, even the demise of promising startups. This sparked a desire in him to create a solution that would empower these entrepreneurs to accelerate the adoption of their innovations. His association with Bharat Innovation Fund – a deep-tech, sector-agnostic fund headquartered at IIM Ahmedabad, led him to believe in the power of technology to solve large problems.

Driven by this vision, he joined forces with three other co-founders to launch ThinkAg, a non-profit platform bringing together various stakeholders in the agri-industry. The goal was to assist startups in advancing their ideas while facilitating early customer contacts with large B2B customers and industry partners. ThinkAg quickly became the first agri-food-fintech platform in India, creating a vibrant ecosystem connecting innovators, corporates, incubators, accelerators, investors, farmer producer organizations (FPOs), and other key players.

Hemendra's contributions extended beyond ThinkAg. In 2019, he was invited by the Federation of Indian Chambers of Commerce and Industry (FICCI) to chair a task force on agri-startups. Under his leadership, the task committee launched initiatives to create a more enabling environment for agri-startups, including workshops, training programmes, and facilitating connections between startups and industry partners.

However, just as these initiatives were gaining momentum, the world was hit by the COVID-19 pandemic. The lockdown measures severely impacted the agricultural sector, disrupting supply chains and threatening the livelihoods of millions. In this moment of crisis, Hemendra in his personal capacity as well as through his association with both ThinkAg and FICCI supported many startups and organizations to keep the operations running.

ThinkAg swiftly transitioned to virtual operations, successfully running 50 to 60 online events that brought together diverse stakeholders to discuss crucial topics such as the digitization of the agri-value chain and agri-fintech potential. Meanwhile, at FICCI, Hemendra and his colleagues played a crucial role in navigating supply chain issues through industry connects and policy support. They rallied individuals from different facets of the agri-industry to work towards a common goal, facilitating collaborations between farmer groups, agribusinesses, logistics providers, and government agencies.

Through these tireless efforts, Hemendra and his teams were able to help many agribusinesses and agri-startups navigate the challenges of functioning during the pandemic and maintain the flow of essential goods and services. Their work not only mitigated the immediate impact of the crisis but also laid the foundation for future collaborations among startups and industry partners.

This experience served as a powerful reminder of the importance of collaboration in the agri-industry. Hemendra realized that the spirit of unity and shared purpose that emerged in response to the crisis was essential for the industry to reach new heights and overcome future challenges.

His impact on the agri-industry extends far beyond these organizations. Currently serving as a mentor, adviser, or board member for dozens of enterprises and startups, Hemendra shares with them his wealth of knowledge and experience to guide them towards success. In the past eight years alone, he has worked with more than 300 agri-related firms, leaving an indelible mark on the industry.

Hemendra has authored numerous articles on agri-tech and entrepreneurship, which have been published in leading print and digital media. In 2017, he penned a visionary article on the topic of 'AgriStack' titled 'How technology can drive change in Indian agriculture' for the YourStory website, which could prove to be one of his most significant contributions to the sector. The piece outlined his concept of building a public digital platform that would revolutionize access to crucial agricultural data. At its core, Hemendra's concept of AgriStack comprised three key components: FarmerStack (containing farmer details), FarmStack (providing farm location and farm size data), and CropStack (information on crops grown on the farm).

This public stack promises to dramatically reduce the time taken and transactional cost for innovations to reach farmers, enabling almost instant and targeted access to millions of agricultural households. Its potential applications are vast, from helping startups and industry identify and reach specific farmers to assisting banks in assessing creditworthiness and giving crop loans, as well as enabling governments to design more effective welfare schemes.

The vision of AgriStack quickly gained traction within the industry. The government has recognized its potential and has sought his guidance in developing digital public goods from time to time. Several state governments have also expressed interest in building their own digital stacks in agriculture, with Hemendra working on and assisting in some of these efforts.

Today, he firmly believes that the Digital AgriStack could become the UPI of agriculture, providing a powerful data repository for addressing farmers' challenges related to market access, credit, insurance, and more. The impact is already visible in states like Uttar Pradesh, where data for 20 crops across 16 million hectare is being captured at the gram panchayat level every 15 days – likely the world's first crop stack of this scale. Interestingly, this crop stack is built by four agri-tech startups with remote sensing capabilities.

Hemendra also sees potential for many other digital stacks in agriculture, such as soil stack, water stack, weather stack, and dairy stack, which have transformative potential for the sector. He believes that many of these stacks will come alive in the next ten years, driving the disruption in the sector that it has never seen before.

For Hemendra, seeing his conceptual vision become a reality within eight years is a dream come true. Alongside his work on Digital Public Goods, he has also been actively collaborating with multilateral organizations like the International Finance Corporation to bring numerous agri-tech innovations to farmers in India, in partnership with state governments such as those of Uttar Pradesh, Madhya Pradesh, and Telangana. These efforts aim to leverage the power of technology to improve the lives of farmers and enhance the efficiency of the agricultural sector in these states.

In addition to his efforts in India, Hemendra collaborated with an Australia-based company called Beanstalk to research and gather information regarding the adoption of digital agriculture tools across developing countries. The study, which is the most comprehensive of its kind, covered four regions: South Asia, Latin America, Southeast Asia, and Africa.

The research sought to understand how farmers are embracing digital tools, the impact these tools have on their lives, and whether they are helping to address climate-related challenges. The study also examined the gender neutrality of these solutions, the performance of agri-tech startups, and how they can be better supported. The insights gleaned from this comprehensive research will be invaluable in shaping future interventions and policies designed to improve the lives of farmers in these regions.

Inspired by the findings of this study, Hemendra has developed a keen interest in exploring global opportunities for India within the agri-tech industry. He draws parallels between the current state of Indian agri-tech and the early days of the Indian IT sector in the late 1980s and early 1990s, when the country was gradually transforming into a global outsourcing hub. In his view, many agri-tech solutions developed in India hold the potential to be equally relevant and impactful in other countries facing similar agricultural challenges, such as Kenya, Tanzania, Indonesia, the Philippines, and nations across Latin America.

As a developing country, India has made significant strides in agri-tech, positioning itself as a leader in this field. Recognizing the unique opportunity to share its expertise and solutions with other nations, Hemendra is actively engaging with various organizations and working towards establishing a global corridor for agri-tech innovations, facilitating the exchange of knowledge and technology across borders.

The Indian government has acknowledged the crucial role of agri-tech startups in driving the sector's growth and has taken steps to support their development. In 2023, the government launched a ₹500-crore acceleration fund for agri-startups, with Hemendra serving as a board member. He is tirelessly working to attract more capital to the Indian agri-tech sector, setting an ambitious target of securing at least $1 billion in annual investments. Additionally, his focus has shifted towards nurturing sustainable solutions and businesses, helping many climate-tech startups that can drive climate risk mitigation and promote sustainable agricultural practices.

Hemendra's embrace of entrepreneurship has profoundly shaped both his professional life and personal outlook. He likens the experience to attaining nirvana, which brings with it a sense of humility, groundedness, and an ability to effect change. This journey has pushed him to explore the depths of his potential, uncovering capabilities he might never have discovered in a traditional career.

As an entrepreneur, he thrives on challenges, approaching each obstacle with a sense of ownership and accountability. He finds the weight of responsibility liberating, allowing him to focus on driving progress and finding solutions. This entrepreneurial mindset, characterized by a relentless pursuit of growth and innovation, fosters a natural positivity that propels him forward.

Throughout his career, his interactions with farmers in Indian villages have been transformative, providing him with invaluable insights into the agricultural sector's challenges and opportunities. These experiences have shaped his vision for the future of agriculture in India, fuelling his determination to drive positive change through entrepreneurship and technology.

This determination is reflected in Hemendra's golden rule: 'Never say no to an entrepreneur. Assist them in any way you can.' His unwavering commitment to supporting agri-tech innovators has led many to unofficially refer to him as the godfather of the Indian agri-tech industry. His guidance and mentorship have profoundly impacted the sector, making him a beacon of hope and inspiration for those working to revolutionize agriculture through technology.

As Hemendra continues his dedication to the growth and development of the Indian agri-industry, his influence extends far beyond the present. His vision for a more connected, efficient, and prosperous agricultural landscape is steadily becoming a reality. With Hemendra Mathur championing innovation and nurturing the next generation of agri-tech entrepreneurs, Indian agriculture stands poised for a transformative leap into a technologically advanced and sustainable future.

# 6

## The Yolk Reformer:
## Abhishek Negi, Eggoz

*F*ROM THE PRESTIGIOUS HALLS *of IIT to the gritty reality of poultry farms, Abhishek Negi's journey defies conventional wisdom. Abandoning a cushy corporate job, he plunged into entrepreneurship with a failed cab aggregator, only to shock everyone by diving into the notoriously challenging world of agriculture. Failure after failure in various farming ventures would have broken most, but for Abhishek, each setback was fuel for his resilience. His relentless pursuit culminated in Eggoz, revolutionizing not just an industry but also a nation's perception of a nutritional powerhouse. Beyond elevating poultry farmers' lives, Abhishek's true triumph lies in transforming millions of consumers into informed egg enthusiasts, making quality eggs a conscious choice rather than an afterthought. In Abhishek's story, we find the essence of entrepreneurship – not in avoiding failure, but in the audacity to rise, reinvent, and reform, even when the odds are stacked against you.*

The path to becoming an agri-tech entrepreneur began in the crucible of adversity for Abhishek Negi. Born into a military family, his childhood was shaped by the realities of conflict. In 2002, as tensions between India and Pakistan reached a fever pitch, eleven-year-old Abhishek, along with his younger sister and mother, found themselves confined to their home in Akhnoor, Jammu and Kashmir, where his father was stationed. The constant sounds of gunfire and explosions became the soundtrack of their days, while air-raid sirens punctuated their nights.

In this atmosphere of fear and uncertainty, the young boy witnessed something remarkable: the power of community. Neighbours banded together, sharing resources and offering comfort in the darkest hours. This experience instilled in him a deep appreciation of life's simple pleasures and the importance of human connection – lessons that would prove invaluable in his future endeavours.

As Abhishek grew older, his academic prowess became evident, and he set his sights on the prestigious IITs. When his father was transferred to Kota, Rajasthan, a city renowned for its IIT-JEE coaching institutes, he seized the opportunity to join Bansal Classes, a coaching centre, and threw himself into his studies. This relentless focus paid off in 2009 when he secured an impressive all-India rank of 689 in the IIT-JEE, earning him a spot in the electrical engineering programme at IIT Kharagpur.

At the institute, the ambitious student dove headfirst into campus life. He participated in various sports and cultural programmes, but it was the annual Illumination Competition that truly captured his imagination. This event, where students created massive, elaborate structures adorned with thousands of earthen lamps, pushed him and his peers to their limits.

For two gruelling months, Abhishek and his hostel mates worked tirelessly, often through the night, to construct their *chattai* – a bamboo structure that would serve as the canvas for their illumination masterpiece. The competition taught them invaluable lessons in teamwork, resource management, and the power of shared goals. The young engineer learned that with motivation and collaboration, people could achieve extraordinary things, even with limited resources.

Driven by a desire to make a larger impact on campus, he ran for the position of general secretary of sports in his second year. This campaign tested his resolve in ways he hadn't anticipated. During a crucial Q&A session, he faced tough questions about his past performance, which threatened to derail his campaign. In a moment of genuineness, he chose honesty over political manoeuvring. Acknowledging his shortcomings, Negi explained the lessons he had learned and outlined how these experiences would make him a better leader.

This display of authenticity resonated deeply with his peers, earning him their trust and, ultimately, their votes. As general secretary, Negi successfully organized the 47th inter-IIT sports meet, an experience that honed his leadership skills, helped him overcome his fear of public speaking, and reinforced the importance of genuine interactions in building trust.

Armed with these valuable experiences and a strong technical background from IIT Kharagpur, the young graduate embarked on his professional journey by joining Vodafone India as a management trainee. This role was part of a groundbreaking initiative by the company to hire IIT graduates with technical backgrounds, rather than solely focusing on IIT MBAs.

Negi's first assignment thrust him into unfamiliar territory – selling SIM cards. For an engineer accustomed to solving technical problems, persuading people to make purchases proved daunting. Despite falling short of his monthly targets and facing frequent rejection, he embraced the challenge, recognizing it as an opportunity to develop crucial communication and interpersonal skills.

After this eye-opening experience in sales, the ambitious trainee transitioned to the engineering department, where he encountered a significant problem: poor network quality for road travellers in Guwahati. Negi and his team worked diligently to find a solution, trying various strategies to improve network quality, but initially made little progress. As days turned into weeks with no apparent breakthrough, he found himself at a crossroads. His manager and colleagues suggested abandoning the project and moving on to something else, but his instincts told him that a solution was within reach.

Taking a proactive approach, the engineer began visiting nearby mobile phone towers to gather more information and potentially uncover a solution. What he discovered surprised him: many of the towers were in a state of disrepair, covered in dust and cobwebs, with misaligned primary antennae and loose connectors. Although uncertain that cleaning the towers would resolve the issue, Negi decided it was worth a try.

To test his hypothesis, he focused on one tower as a pilot project. Climbing the structure, he cleaned it thoroughly and fixed the connectors and antenna. After a week of monitoring the tower's statistics, Negi and his team noticed a significant decrease in the call drop rate in the area. Encouraged by this success, they set out to repair and clean the remaining 1,200 towers over the next month.

Their efforts yielded remarkable results. Analyzing the tower data after completing the repairs, they found that the overall network quality metrics had improved significantly. The young engineer was thrilled to have found a solution that not only benefitted his company but also improved the experience of countless mobile phone users in the region. This success boosted his confidence and reinforced his belief in the power of persistence and in thinking outside the box when faced with a challenge.

Impressed by Negi's initiative and problem-solving skills, his superiors at Vodafone India decided to transfer him to the company's office in Kolkata after just six months of service in Guwahati. This move marked the beginning of a new chapter in his life, one that would ultimately lead him down the path of entrepreneurship and shape his future in ways he had never imagined.

During his weekends in Kolkata, the young professional often found himself drawn back to his alma mater, IIT Kharagpur, where many of his classmates were still enrolled in five-year integrated programmes. These visits provided him with an opportunity to reconnect with old friends and engage in stimulating discussions about the burgeoning startup industry in India. It was during one of these conversations that Negi recalled an idea he had been contemplating since his college days.

IIT Kharagpur's remote location, more than two hours away from the nearest city of Kolkata, presented a unique challenge for students seeking transportation. The only options available were by rail, in overcrowded and unreserved train coaches, or by prohibitively expensive taxis that charged round-trip fares, sometimes consuming a student's entire monthly budget. Negi, along with his friends Siddhant Matre and Ashish Rajput, believed that there was an untapped market for a service that could solve this problem by offering one-way taxi fares.

They recognized the immediate need for such a service and envisioned a solution that would involve tracking fleets of dedicated out-of-town travel cars in multiple cities and utilizing an intelligent algorithm to match user demand with available rides. This approach would not only allow drivers to earn a decent living but also enable customers to pay only one-way fares, making intercity travel more accessible and affordable.

As Negi delved deeper into the potential of this idea, he began to reflect on his own experiences and the lessons he had learned throughout his life. From childhood, he had always been driven by a desire to make a positive impact on the lives of others. He realized that this transportation solution could be more than just a business venture; it could be a way to empower people and create lasting change.

Pursuing this dream would mean leaving behind the stability and growth opportunities at Vodafone, where he had built a reputation for solving critical problems. But his youthful energy and optimism propelled him forward. With a resolute heart, he bid farewell to the corporate world and embarked on this new journey in 2014 with Siddhant and Ashish, all three pooling their savings to fund their venture, which they named Roder.

The next challenge was deciding where to launch their business. After careful consideration, they chose the National Capital Region (NCR) for its proximity to numerous cities, the high volume of people entering and exiting the region daily, and the lack of significant competitors at the time. Uber had not yet entered the Indian market, and Ola had only recently begun providing airport transportation in NCR, giving Roder a unique opportunity to establish itself in the market.

Before launching the service, Negi took a hands-on approach to understand their market. For a month, he drove his uncle's car across NCR, picking up and dropping off customers and gaining invaluable insights into the needs of both drivers and passengers.

Bolstered by this first-hand experience, the team launched their outstation cab aggregator service with great enthusiasm. Their revenue model was straightforward: they operated as a 'digital cab aggregator for outstation travel', essentially functioning as an 'Uber for long-distance trips'. Roder would connect passengers with drivers for intercity travel, taking a commission from each booking.

The company slowly began to establish its presence in the NCR market, but it wasn't without challenges. The team soon realized that the taxi industry, by its very nature, required significant investments of both time and money to build a sustainable ecosystem of drivers. This constant need for funding proved to be a major challenge for the young entrepreneurs.

Despite their best efforts to live frugally and allocate most of their funds to the company, Negi and his co-founders found that their limited savings were insufficient to scale the business as quickly as they had hoped. The team began to pitch their idea to potential investors. After a year of multiple rejections, they finally received a glimmer of hope in the form of a $240,000 initial investment in January 2016.

This funding allowed them to hire employees, redesign their mobile application, and expand their service to additional locations. Over time, Roder built a presence in over 30 locations and achieved an impressive 30 per cent month-on-month revenue growth. However, this rapid expansion also led to a significantly increased need for additional funding, forcing the team to once again seek out investors for the next round of financing.

Initial meetings with investors yielded positive responses, and it seemed that the necessary funding was within reach. However, unknown to Negi and his team, a storm was brewing beneath the surface that would soon threaten the very existence of their company.

In early 2016, two major players in the ride-hailing industry, Uber and Ola, aggressively entered the intercity taxi service market, armed with deep pockets and extensive resources. These companies launched promotional incentives that made investors hesitant to back smaller startups like Roder, fearing for their survival in the face of such fierce competition.

Despite their determination to weather the storm and eventually attract investors, the Roder team was dealt another devastating blow in the form of demonetization.

The economic climate and consumer demand took a significant hit, further compounding the challenges faced by the startup. Their investor urged them to seek a buyer for the business, but the prevailing market conditions made it nearly impossible to find a suitable acquirer. Without sufficient funds to continue operations, Negi and his team were left with no choice but to shut down the company.

As the reality of their company's collapse settled in, the co-founders found themselves engulfed in a darkness that seemed to stretch endlessly before them. The decision to shut down Roder was an emotional one, considering the time, energy, and soul they had poured into building the company. In the aftermath of the shutdown, Negi took a three-month hiatus to navigate the psychological toll it had taken on him. He found himself at a crossroads, uncertain of his future path. The allure of joining another company and settling into a stable job had lost its lustre, and the thought of starting anew seemed daunting. As his classmates forged ahead in their careers, the young entrepreneur felt as though he was standing still, enveloped by a suffocating darkness.

During this period of introspection, Negi maintained contact with two of his friends from IIT Kharagpur, Uttam Kumar and Aditya Singh. Uttam, a seasoned entrepreneur who had previously founded and sold a company, was now on the lookout for a new business opportunity. Aditya, hailing from a farming background, harboured a deep passion for the agricultural sector, believing there were untapped opportunities waiting to be explored and that farmers were not receiving their fair share of profits.

As Uttam and Aditya discussed their ideas and aspirations, Negi found himself drawn into their conversations, unconsciously becoming more involved with each passing day. Despite his initial hesitation and the wound of his past failure still fresh, he felt a glimmer of excitement at the prospect of embarking on a new venture with his friends.

The young entrepreneur knew that starting another business would require a special kind of courage, the kind that would allow him to jump into another well after being hurt both physically and emotionally in his earlier endeavour. It would be a leap of faith, a test of his resilience and determination. And as he listened to his friends' plans, he felt a spark ignite within him, and he felt a renewed sense of purpose that had been absent in the months following Roder's shutdown.

At Aditya's urging, the three of them conducted a preliminary study of the agriculture industry. Their research led them to conclude that there was potential for profit by establishing a direct link between farmers and consumers, eliminating intermediaries and ensuring that farmers received a larger share of the proceeds from the sale of their produce.

Despite their lack of experience in farming or related activities, the group decided to pursue this idea, driven by their shared desire to positively impact the lives of farmers. However, as they soon discovered, executing their plan would prove to be far more challenging than they had initially anticipated.

In 2017, the friends started their new business in one of Mumbai's local neighbourhoods, delivering fruits and vegetables to domestic customers on their scooters daily. However, within just a few days, they realized that despite working hard and putting in long hours, they were unable to generate sufficient profits to sustain their operations. This harsh reality forced them to question their earlier decision and approach.

After careful deliberation, they concluded that they lacked extensive knowledge of farming and the intricacies of the supply chain, and this was a significant obstacle to their success. Negi proposed that the best way to acquire this knowledge would be to relocate their business to a rural area, where they could immerse themselves in the lives of the farmers they aimed to help. This strategic move would allow them to gain valuable insights and a deeper understanding of the challenges faced by the farming community.

With this new plan in mind, they decided to move to Bihar Sharif, Uttam's hometown in Bihar, where they could leverage his local connections to gain a foothold in the community. Negi, who was now penniless, found a silver lining in their relocation. The lower cost of living in Bihar Sharif meant that living there would be less of a financial burden, allowing him and his friends to focus their resources on building their business.

The group rented a small house and began immersing themselves in the daily lives of the farmers. They spent countless hours in the agricultural fields and local trading markets, known as mandis, observing and learning about the challenges facing the farming community.

Eager to put their new-found knowledge to use, they partnered with farmers in the Nalanda district of Bihar, collecting their produce and trading it in the local mandis. However, after two months of trading, they encountered a new set of obstacles that threatened to derail their progress. The need for warehousing, logistics assistance, and scalable technology proved to be a working capital nightmare. Additionally, the payment structure of the industry required them to pay the farmers on the day of produce collection, while they would only receive payment from their buyers a week later. This meant that to increase their business tenfold, they would need to multiply their initial investment by the same factor. Negi, still haunted by the financial challenges of his previous venture, grew cautious about proceeding down this path.

The team realized that they needed to pivot their idea. It was in this moment of uncertainty that a critical observation was made. The small landholdings in Bihar made it nearly impossible for farmers to afford modern harvesting equipment, forcing them to rely on manual labour. This reliance often led to decreased quality of produce, lower yields, and increased wastage – a problem that the team realized held the key to their next move.

Inspired by this realization, the team devised a new plan. They would offer a 'farming-as-a-service' model, purchasing a harvester and renting it out to small farmers who could not afford to buy their own equipment. The group meticulously planned their pricing structure and projected profitability, even going so far as to teach themselves how to operate the harvester to ensure the success of their venture.

To test the market response, they began advertising their services and offered subsidized harvesting to a select group of farmers, hoping this would attract more customers willing to pay for their services. However, despite their efforts and the expansion of their outreach, they were met with a surprising lack of interest. The farmers simply did not take up their paid services.

Discouraged but determined to understand the root cause of the problem, the team realized that the issue was not with the quality of their service but rather with the financial constraints faced by the small farmers in the region. These farmers simply did not have the disposable income to pay for such services. This realization led them to conclude that the best way to help these farmers would be to find a business model that paid the farmers directly rather than expecting them to pay for services.

Confronted with the harsh truth that their farming-as-a-service model was not viable in the current climate, the friends found themselves at yet another pivotal juncture. They had poured their hearts and a significant portion of their financial resources into acquiring harvesting equipment, and the looming possibility of yet another failure weighed heavily on their minds. It's a question many entrepreneurs grapple with: how many times can one pick oneself up from defeat and find the courage to take another leap of faith? Each attempt had left them battered and bruised, but their indomitable spirit and the invaluable knowledge they had gained refused to let them retreat to the familiarity of the corporate world. They remained steadfast in their commitment to finding a path that would allow them to generate revenue while uplifting the lives of local farmers.

As they pondered their next move, a serendipitous visit to a local mandi in Bihar Sharif in late 2017 provided the spark of inspiration they so desperately needed. The group couldn't help but notice the endless stream of trucks arriving at the mandi, each laden with a staggering two and a half lakh eggs. These trucks, they discovered, originated primarily from the distant southern states of Andhra Pradesh, Telangana, and Tamil Nadu. The sheer volume of eggs consumed daily in this small town – a mind-boggling 1 million eggs per day – ignited a spark of curiosity in them. The fact that nearly all of these eggs were brought in from outside Bihar only added fuel to the fire of their inquisitiveness.

Intrigued by this discovery, Negi and his team embarked on a journey to unravel the mysteries of the poultry supply chain in Bihar. As they immersed themselves in the study of the industry, they began to uncover a world of untapped potential and hidden opportunities. Their research revealed a startling disparity: in 2017, while Bihar consumed more than 3 crore eggs daily, production in the state was barely 15 lakh. This glaring gap highlighted a myriad of supply chain issues that needed to be addressed.

The team discovered that the majority of egg production took place in South India, while the bulk of egg consumption occurred in the northern region. This geographical mismatch meant eggs had to endure a gruelling ten-day journey in open trucks to reach their destinations in the north. The long transit time not only compromised the quality of the eggs, causing them to lose their essential nutrients, but also facilitated the growth of harmful bacteria like salmonella. Moreover, with 97 per cent of eggs in India being consumed loose, consumers had no way of tracking the origin or determining the quality and freshness of the eggs they purchased.

As the friends delved deeper into their research, they realized that the market potential for healthier eggs in India was immense. Nearly 95 per cent of the population consumed eggs, with the Indian Council of Medical Research (ICMR) recommending a minimum consumption of 181 eggs per capita per year. However, in 2017, actual consumption of eggs stood at a mere 69 per year per person, in stark contrast to the 240 eggs consumed by adults in most developed countries. This discrepancy underscored the urgent need for healthier eggs in large quantities.

Determined to dig deeper, the team visited poultry farms across the region. What they witnessed was eye-opening: poor health and safety standards, coupled with rampant use of antibiotics, hormones, and concentrates to boost productivity. The team learned that poultry farmers often resorted to these practices to ensure a regular income, as the commodity nature of eggs made it challenging for them to invest in proper farm maintenance or organic, healthy feed.

It became evident that the current state of the poultry industry in India not only affected the quality of eggs but also had a significant impact on the lives of poultry farmers. The farmers were caught in a vicious cycle, where the pressure to maintain a steady income forced them to compromise on the health and well-being of their birds.

This, in turn, led to the production of substandard eggs and further perpetuated the farmers' financial struggles.

Furthermore, the lack of transparency in the supply chain and the absence of a strong brand identity for eggs meant that farmers had little bargaining power and were often at the mercy of market fluctuations. This made it difficult for them to secure a fair price for their produce and invest in improving their farming practices.

Rather than viewing these challenges as obstacles, Negi, Uttam, and Aditya saw them as a golden opportunity to create meaningful change. They envisioned a business that would revolutionize the way people consumed eggs, providing them with a nutritious and fresh alternative to the sub-par options currently available. They wanted to build a company that was more than just a profitable venture – one that would empower farmers rather than exploit them, and one that could scale sustainably without the constant need for substantial capital injections.

There was just one tiny hiccup: none of the three knew the first thing about poultry farming. But they were determined not to let this obstacle stand in their way. With the same tenacity and hunger for knowledge that had propelled them through their engineering studies at IIT Kharagpur, they immersed themselves in the world of poultry science. They devoured research papers, sought out industry experts, and attended every workshop they could find.

As they delved deeper into the intricacies of egg production, a crucial realization dawned on them: the key to producing high-quality eggs lay not in post-harvesting techniques but in the pre-harvesting practices and methods employed by farmers. Armed with this insight, they set out to identify the specific procedures and scientific practices that could help increase productivity and yield fresh, nutrient-rich eggs.

However, Negi and his team knew that simply possessing this theoretical knowledge would not be enough to gain the trust and respect of the farmers they hoped to work with. As young, college-educated entrepreneurs with no practical experience in poultry farming, they risked being seen as outsiders who couldn't possibly understand the real challenges faced by those in the field.

They realized that the only way to bridge this gap was to gain hands-on experience themselves. They considered two options: either work on an existing farm that already employed advanced scientific techniques, or build their own farm from scratch to operate and experiment on. While working on an established farm might have been the easier path, they knew it would limit their ability to innovate, test new ideas, and rapidly iterate based on their findings. Building their own farm, on the other hand, would give them the freedom and flexibility they needed to truly push the boundaries of what was possible.

But building a farm from the ground up was no small feat, and it required substantial financial investment. The friends had already poured their hearts and souls into their previous entrepreneurial ventures, leaving their personal resources severely depleted. Taking on this new challenge meant risking everything they had left, with no guarantee of success. The spectre of their past failures loomed large, and many in their position would have baulked at the idea of jumping back into the fray so soon. Yet, the team made the bold decision to approach angel investors and successfully secured ₹80 lakh to build their dream farm.

With that money, and using their engineering knowledge, Negi and his team meticulously designed and built an efficient farm in 2017. They chose to establish their first farm in Kanchanpur village, Bihar Sharif, Bihar, incorporating the best practices they had learned during their research. The first challenge was to select the right chicks. Drawing upon the knowledge gained from their research, they meticulously chose the healthiest specimens from reputable poultry breeding companies, ensuring a strong foundation for their flock. However, the process was far from simple. Long hours were spent examining each chick, consulting with experienced breeders, and making difficult decisions when some didn't meet their stringent criteria.

With the chicks selected, the team turned their attention to the daily operations of the farm. Feeding schedules had to be established, and the nutritional requirements of the growing birds carefully monitored. Vaccinations were administered with precision, as the engineers-turned-farmers quickly learned the importance of disease prevention in the confined quarters of the poultry sheds.

Cleaning the sheds proved to be a daunting task, far removed from their work in the air-conditioned labs and offices they were accustomed to. The pungent odour of ammonia from the bird droppings filled the air, and the physical labour of mucking out the sheds left them exhausted and sore. Yet they persevered, knowing that maintaining a hygienic environment was crucial to the health and productivity of their flock.

Monitoring the conditions within the sheds was a delicate task, as even small fluctuations in temperature or humidity could have devastating effects on the birds. Drawing upon their engineering backgrounds, Negi, Aditya, and Uttam designed an efficient climate-control system, utilizing sensors and automated ventilation to maintain optimal conditions. They spent long nights huddled over schematics and wiring diagrams, using the skills honed over their academic and professional lives to build a state-of-the-art facility.

But the challenges didn't end there. Power outages were a common occurrence in the rural area, threatening the delicate balance they had worked so hard to achieve. The team found themselves constantly on call, rushing to the farm at odd hours to address emergencies and ensure the well-being of their birds. Sleep became a luxury as they juggled the demands of their new life, struggling to maintain their own physical and mental well-being amidst the relentless pressure.

There were moments of doubt, when the sheer magnitude of their undertaking seemed overwhelming. The long days, the physical toll it took on them, and the financial risks weighed heavily on their minds. But in those moments, they drew strength from each other and from their unwavering belief in their mission. As the weeks turned into months, their hard work began to bear fruit. The chicks grew into healthy, productive birds, and the quality of the eggs they produced exceeded even their own expectations, marking a significant milestone in their entrepreneurial journey.

As Negi and his team continued to operate their farm in Bihar Sharif, an exciting opportunity presented itself. A farm in Madhya Pradesh, complete with 30,000 live birds, became available for purchase. The team recognized the opportunity to expand their operations and gain valuable experience in a new location. While their ultimate goal was to establish a consumer-facing egg business, they understood that running their own farms would provide them with the hands-on knowledge and credibility needed to win the trust of local farmers, who would eventually become their suppliers.

Seizing the opportunity, the team took over the farm in Madhya Pradesh, eager to apply the lessons they had learned in Bihar Sharif. However, they quickly realized that the climatic conditions in this new location posed unique challenges, rendering their previous knowledge insufficient. They dove headfirst into extensive research again, poring over thousands of scholarly materials and engaging with veterinarians and nutritionists to fine-tune their approach. Through a series of short trials, they determined the optimal feed formulation and feeding rates for the specific climatic conditions and bird ages. Their efforts paid off, and within months, Negi and his team had successfully cracked the code, discovering the ideal conditions for shed illumination, temperature, noise levels, cleanliness, and air composition.

The results were nothing short of remarkable. The team, who had started with no prior experience in poultry farming, found themselves successfully operating two farms with productivity rates exceeding 90 per cent.

Buoyed by their success, the team was eager to share their expertise with other farmers in the region. They believed that by implementing their scientifically proven techniques, they could help these farmers improve their productivity and, consequently, their livelihoods. Moreover, by partnering with local farmers, they could secure the large volume of egg supply needed to launch their consumer-facing business.

However, when they approached the farmers with their proposal, they were met with scepticism and resistance. Many farmers were hesitant to entrust the care of their birds to outsiders, fearing that any changes to their traditional methods could disrupt the delicate balance of their operations and potentially harm the health and productivity of their birds.

But Negi and his team persevered. They continued to reach out to farmers in the area, patiently explaining to them the benefits of their scientific approach and the potential for increased productivity and profitability. Gradually, their perseverance paid off, and a small group of five farmers who were struggling to manage their own poultry farms reached out to them for assistance.

Seizing this opportunity, the team worked closely with these farmers, implementing their proven techniques and carefully monitoring the results. To their delight, the farms under their management quickly began to thrive, with productivity rates soaring and the quality of the eggs improving significantly. Word of their success spread rapidly throughout the farming community, and soon more farmers were seeking their services, eager to adopt the same practices and reap the benefits for themselves.

Recognizing the transformative potential of their methodologies, Negi and his co-founders began to explore new ways to expand their network and make an even greater impact in the poultry industry. They realized that by encouraging more individuals to start their own poultry farms and providing them with the necessary support and guidance, they could not only increase the overall production of high-quality eggs but also create new opportunities for entrepreneurship and economic growth in rural communities.

However, they soon discovered that one of the most significant barriers to entry for potential poultry farmers was the requirement for a minimum farm size of 12,000 birds. This posed a considerable financial and logistical obstacle for many prospective farmers, limiting the pool of entrepreneurs who could participate in the industry.

This size requirement stemmed from the typical elevated structure of poultry farms in India, where birds are placed 15 feet above the ground to accommodate their litter. The elevated construction necessitated the use of a large quantity of construction materials at the foundation, driving up the cost of the shed and making it financially viable only for larger farms. Smaller sheds faced problems arising from the accumulation of bird droppings and excrement on the ground, which could take up significant space and create an unsanitary environment for the birds. Inadequate waste removal could also pollute the air and pose health hazards.

While larger farms could invest in equipment like conveyors and other machines to collect litter and improve air quality, these solutions were often not feasible for smaller farms, leading to reduced egg production, diminished quality, and financial losses. To overcome this size limitation, Negi and his team put their engineering skills to work, developing a small, modular design for 2,000-square-foot sheds that placed the birds at ground level. They also created inexpensive conveyor belts for removing the litter. These belts could be operated manually or semi-automatically, allowing farmers or their families to control them with ease.

These innovations significantly reduced the investment required to build and operate a poultry farm, bringing costs down from ₹80 lakh to as low as ₹1.5 lakh, making poultry farming more accessible to small farmers. With the support of the Gates Foundation, which sponsored four of these poultry farm models, and the ICICI Foundation, the team was able to open several more farms. These smaller farms proved to be highly profitable and contributed to increased egg production.

The success of these farms had a ripple effect. The reduced barriers to entry and the promise of a sustainable livelihood attracted a new generation of entrepreneurs from the surrounding villages. Young, ambitious individuals eagerly embraced the opportunity to start their own poultry farms, armed with the knowledge and support provided by Negi and his team. These budding entrepreneurs not only earned respectable incomes but also contributed to the economic growth and development of their communities.

As the number of farms under their guidance grew, so did the production of high-quality eggs. The team found themselves facing a new challenge: building an efficient and effective network to sell their products. They immersed themselves in the task of creating a robust distribution system, forging relationships with retailers and wholesalers across the region. At this stage, the idea of branding eggs had not yet crossed their minds. Negi and his co-founders remained laser-focused on tackling the challenges that plagued the back end of the poultry supply chain, from improving farm management practices to streamlining logistics and distribution. However, an event involving a kathi roll vendor in the Bihar Sharif market would plant the seed for the idea of branding their eggs.

It was a seemingly ordinary day at the market when the team noticed something peculiar. One of their regular customers, a kathi roll vendor, had experienced a sudden surge in demand for egg rolls made with their fresh eggs. Intrigued by this development, they approached the vendor to learn more.

The vendor revealed that customers had been raving about the superior taste and quality of the egg rolls, attributing the difference to the eggs supplied by Negi's team. This revelation sparked an idea: if their eggs could make such a noticeable impact on a single vendor's business, perhaps there was potential to charge a premium for their eggs, which already stood out in the market.

Eager to test this hypothesis, the team decided to implement a price increase for their eggs, confident that the superior quality would justify the adjustment. To their surprise, the vendor initially declined to buy their eggs, opting to source cheaper eggs from other suppliers to maintain his profit margins. However, the vendor soon discovered that his customers were not satisfied with the change in egg quality. Sales began to decline, and customers expressed their disappointment in the deterioration of the once-beloved egg kathi rolls. The vendor reluctantly returned to Negi's team, agreeing to pay the premium for their superior eggs. The incident served as a powerful testament to the value of their product and marked the first time the team seriously considered the potential for branding their eggs to differentiate themselves from competitors.

Despite this revelation, the team did not actively pursue the creation of their own brand of eggs. However, the onset of the COVID-19 pandemic and the subsequent lockdown in April 2020 presented them with an unprecedented challenge that would ultimately accelerate their branding plans.

During India's initial COVID-19 lockdown, a rumour began circulating that consumption of eggs could cause COVID-19. This misinformation led to a sharp decline in egg demand, causing egg prices to plummet. As a result, farmers across the country, including Negi and his team, suffered severe losses. The financial strain of the situation forced them to make the difficult decision to cull one of their flocks, as they could no longer afford to feed the birds.

This experience served as a wake-up call for the team, highlighting the vulnerability of relying solely on commodity pricing. They realized that to protect themselves from such market fluctuations and ensure the sustainability of their business, they needed to establish their own brand and pricing system.

In May 2020, just a month after the lockdown began, they launched Eggoz, their branded eggs, in Gurgaon. The team hit the ground running, visiting apartment complexes and setting up demo stalls to introduce their products to potential customers. As people sampled the eggs, they were impressed by their freshness, taste, and pleasant aroma when cooked. Word spread quickly, and more residents flocked to the stalls, eager to experience the difference for themselves.

The founders took this opportunity to educate customers about the importance of checking for freshness, cleanliness, and production dates when purchasing eggs. The positive response they received not only validated their efforts but also bolstered their confidence in the Eggoz brand.

Buoyed by this success, the team approached local grocery stores, convincing them to stock Eggoz eggs for customers who had enjoyed the demos. They also partnered with online platforms like Instamart, BigBasket, and Zepto, expanding their reach and making their products more accessible to a wider audience.

Through extensive research and on-the-ground experience, Negi and his team discovered that launching their brand in tier 2 cities, where competition was less fierce, would be more effective in building brand awareness and driving sales. Armed with this knowledge, they focused their efforts on establishing Eggoz in these smaller cities.

As Eggoz continued to grow and gain traction, the founders recognized the need for additional funding to fuel their expansion plans. In 2018, they embarked on the challenging journey of seeking investors. Over the course of a year, they attended countless meetings and faced numerous rejections, each one presenting a new hurdle to overcome.

One of the primary reasons for these rejections was that their venture did not have a flashy website or app, which deterred some investors who placed a high value on such features. Others dismissed Eggoz as a traditional farming business, arguing that there was nothing innovative about it and that it lacked the potential for significant growth and impact. Many investors were hesitant to take a risk on what they perceived to be a conventional venture.

Negi frequently encountered the belief that creating a brand in the egg market was an impossible feat. He countered this notion by pointing out that twenty-five years ago, the idea of branding water seemed far-fetched, yet countless water brands exist today. He acknowledged that his company might face challenges in execution or lack certain skills, but he firmly believed that creating a brand in the egg industry was not only possible but inevitable. He cited the United States as an example, where 83 per cent of eggs are branded, in contrast to India's mere 3 per cent brand penetration, dominated by local companies.

The entrepreneur argued that Indian consumers were evolving, and opportunities for branded food products were on the horizon. Some investors, however, refused to invest in Eggoz simply because eggs are not vegetarian. Others raised concerns about the potential impact of bird flu outbreaks on the business.

Negi addressed these concerns head-on, drawing parallels to the risks faced by internet companies in the event of a sudden outage or the imposition of new digital taxes. He acknowledged that bird flu posed a threat but emphasized the importance of implementing biosecurity measures, planning for such occurrences, and taking the necessary precautions. In its first five years of operation, Eggoz faced three cases of avian flu, each resulting in a temporary drop in demand and a three-to-four-week recession. However, the company managed to navigate these challenges and emerge stronger.

Despite the numerous rejections, Negi and his team persevered, continuing to meet with investors until they finally secured a second round of funding with the help of angel investors and some of their college seniors who had established a network of investors.

By 2022, Eggoz had established a strong presence in Delhi, NCR, and Bengaluru. However, the post-COVID era brought new challenges, with investors tightening their purse strings and demanding profitability. Negi realized it was time to consolidate rather than expand – a pivot that would prove to be a defining moment for Eggoz.

The team embarked on a mission to streamline operations. They implemented a series of strategic initiatives, including system efficiency improvements and supply chain optimization. The founders developed a comprehensive enterprise resource planning system and leveraged data and machine learning to predict demand and understand consumer behaviour. These efforts paid off handsomely, with Eggoz witnessing significant traction and improved operational efficiency.

Once profitability was achieved and systems were running smoothly, Eggoz was ready to spread its wings. The company expanded its footprint to cities like Chennai, Mumbai, Hyderabad, Jaipur, and Chandigarh, receiving positive responses that validated its expansion strategy. However, Negi and his team were acutely aware of the challenges that come with operating in a commodity-driven industry like eggs. The constant pressure on margins prompted them to explore new avenues for growth and differentiation.

The solution came in the form of value-added egg products. Drawing inspiration from the thriving value-added segments in the dairy and chicken industries, Eggoz embarked on a journey to create innovative egg-based offerings. Their first attempt, while innovative, didn't gain traction as expected. But this team was no stranger to setbacks. They quickly realized that familiarity was key when introducing new products to consumers.

Taking cues from Maggi's successful launch in the 1980s, which capitalized on Indian masala flavours, they pivoted their approach. They identified momos as a familiar, popular product with the potential for a value-added offering. Eggoz introduced frozen egg momos, which could be easily prepared by steaming or frying, just like regular vegetable or chicken momos. These rapidly gained traction and became a resounding success. Buoyed by this achievement, they expanded their value-added product line, introducing innovative offerings like egg salami, which could be consumed cold after defrosting, and omelette bites, which could be quickly prepared by microwaving.

These products resonated not only with consumers but also with the hotels, restaurants, and cafes/catering (HoReCa) segment. Eggoz's value-added products emerged as a game changer, providing consistent quality, exceptional taste, and improved operational efficiency to an industry long grappling with challenges like inconsistent egg quality and the complexities of handling eggshell waste in commercial kitchens.

The success of Eggoz's value-added products is a testament to the team's passion and innovative spirit. Negi and his partners fostered a culture of ownership and idea-driven innovation, empowering employees to take charge of their respective areas and align their efforts with the company's overarching goals.

Eggoz's core values have been instrumental in its rapid growth, with a strong emphasis on aligning with purpose, maintaining a customer-centric approach, developing sustainable solutions, embracing a pioneering spirit, and encouraging out-of-the-box thinking.

Another critical factor in Eggoz's success has been its strategic approach to hiring. The founders prioritized bringing on board individuals who were not only the right cultural fit but also had deep expertise in specific domains, even if their knowledge surpassed that of the founders. This hiring philosophy contributed significantly to the organization's growth and provided Negi and his co-founders with invaluable insights and opportunities for leadership development.

Negi strongly believes that founders must continuously evolve alongside their company to keep pace with its growth trajectory. He stresses the importance of relentless learning, actively seeking guidance from mentors, engaging with industry experts, and embracing a hands-on approach through iterations and prototyping. By tackling challenges from a fundamental, first-principles perspective, Negi has successfully navigated the complexities of scaling Eggoz.

Central to Eggoz's success was its unwavering commitment to delivering consistently fresh and high-quality eggs. The company adopted a local production and consumption model, emphasizing the importance of delivering eggs that are 24 to 36 hours old. They established state-of-the-art egg processing centres capable of performing 11 quality checks, ensuring that only the finest eggs reached their customers.

Equally important to Eggoz's business model is fair compensation for their farmers. The company seeks to procure eggs at better prices while providing various support services. As a result, partnering farmers earn at least 30 per cent more than under the old system, making for a mutually beneficial relationship that uplifts rural communities.

As Eggoz continues to grow, it faces new challenges and opportunities. The company must navigate the complexities of scaling operations while maintaining the quality and freshness of its produce, which have become synonymous with its brand. It must also continue to innovate in product development, marketing strategies, and farmer partnerships to stay ahead in an increasingly competitive market.

The story of Eggoz also highlights the potential for transformative change in India's agricultural sector. By applying modern technology, scientific practices, and innovative business models to traditional farming, entrepreneurs like Negi are not only creating successful businesses but also contributing to rural development and food security in the country.

As consumers become more health-conscious and the demand for high-quality, traceable food products increases, companies like Eggoz are well-positioned to lead the way in transforming India's food supply chains. Their success could inspire a new generation of agri-entrepreneurs, potentially revolutionizing other segments of the agricultural sector.

# 7

# The Purpose-Driven Founder: Nidhi Pant, S4S Technologies

*THE ENVIRONMENTS OF OUR youth shape us in profound ways, moulding our values, beliefs, and life trajectories. For Nidhi Pant, co-founder of S4S Technologies, this formative environment was defined by her parents' journey from struggle to success and their unwavering commitment to uplifting others. Growing up in a household where personal achievement was intrinsically linked to community betterment, her childhood experiences laid the foundation for her life's purpose. As our early social circles often determine our decision-making processes, this young visionary's immersion in a world of empathy and action sculpted her into a forward-thinking entrepreneur. She has dedicated her life to empowering farmers through innovative agri-tech solutions. Nidhi's story exemplifies how seeds of compassion, when nurtured by experience and fuelled by determination, can blossom into a powerful force for social change.*

Born into a family of scientists, Nidhi's formative years were shaped by the values of hard work, perseverance, and social responsibility. Her father, the youngest of eleven siblings, hailed from a remote Himalayan village in Uttarakhand. He witnessed first-hand the devastating effects of climate change, with floods and landslides forcing mass migrations and turning many villages in his state into ghost towns. Recognizing education as his path to a better life, he relocated to Mumbai, studied diligently, and secured a position as a scientist at the prestigious Bhabha Atomic Research Centre (BARC).

Both Nidhi's father and mother (also a physicist at BARC) shared a deep commitment to making a difference. Despite their demanding careers, they remained actively involved in community service. Regular visits to their ancestral village in the Himalayas became opportunities not just to help relatives but also to work on projects aimed at uplifting those struggling to live in the harsh conditions.

During these visits, the Pant children were encouraged to find ways to help others. Their parents emphasized that true success and happiness lay in using one's skills for the betterment of those less fortunate. This ethos of growing by uplifting others became deeply ingrained in Nidhi and her sister from a young age.

Even during festivals like Diwali and Holi, the focus in the Pant household was on making a difference. Collecting and distributing clothes, arranging food for the needy, and assisting people with disabilities became integral to Nidhi's upbringing. Her parents led by example, demonstrating that every occasion was an opportunity to create a positive impact.

Nidhi's father was particularly passionate about addressing water scarcity in the Himalayas. Having witnessed the effects of contaminated water on his own family and community, he invested time in developing low-cost water purification solutions for the villages. His daughters actively participated in these projects, gaining first-hand experience in using technology to solve real-world problems.

Nidhi's upbringing in Anushakti Nagar, a residential town for BARC employees in Mumbai, exposed her to an environment that fostered academic excellence and a love for science. In this tight-knit community, education took precedence, with children preparing for competitive exams from an early age. She began her IIT preparation while in the tenth grade, immersing herself in the world of science and technology.

Her academic journey took a slight detour when she opted to pursue chemical technology at the University Department of Chemical Technology (UDCT) in Mumbai. As the daughter of two physicists, Nidhi was drawn to chemical technology's unique ability to bridge theory and application. She saw it as a field that could provide deeper insights into the physical and chemical transformations governing nature.

During her time at UDCT, Nidhi discovered a world beyond academics. The university offered a vibrant environment that encouraged students to explore their passions and develop leadership skills. She quickly took on various roles, starting as cultural secretary in her first year, immersing herself in organizing extracurricular activities and community programmes that fostered camaraderie among students.

As she progressed through her degree, Nidhi realized her true passion lay at the intersection of technology and social impact. She actively sought opportunities to engage in social development work, volunteering with non-profit organizations focused on uplifting marginalized communities. Through these experiences, she witnessed the transformative power of technology in improving lives and driving positive change.

While enjoying the technical aspects of her coursework, Nidhi found herself gravitating towards roles involving writing, project management, and team leadership. It was during this time that she first crossed paths with her future co-founders, who were pursuing advanced degrees at UDCT. The university's web platform served as a hub for students to connect and share ideas. Known for her strong communication skills and innovative thinking, Nidhi quickly became a go-to resource for her seniors.

One of these seniors was Vaibhav Tidke, who was working on a fascinating project related to uplifting rural areas. Vaibhav's passion for using science and technology to improve the lives of the people in these communities resonated deeply with Nidhi, who had grown up watching her parents dedicate themselves to similar causes. The two struck up a friendship, often engaging in lengthy discussions about the potential of technology to revolutionize agriculture and empower farmers.

As Nidhi delved deeper into Vaibhav's work, she learned about his plans to start a company called S4S Technologies, which stood for 'Science for Society'. The company's mission was to develop scientific solutions that could improve the lives and livelihoods of rural populations, which aligned perfectly with Nidhi's own values and aspirations.

Inspired by this vision and the potential impact of S4S Technologies, Nidhi began collaborating with Vaibhav and his team on various projects. She contributed her skills in research, writing, and ideation, working closely with her seniors to refine their ideas and bring them to life. Through these interactions, Nidhi formed strong bonds with the individuals who would later become her co-founders in her venture, united by their shared passion for using science and technology to drive social change.

As graduation approached, Nidhi found herself at a crossroads. Her academic excellence, strong communication skills, and leadership experience at UDCT had positioned her as a prime candidate for lucrative corporate roles. Many of her peers were eagerly accepting high-paying jobs in multinational companies, a path that promised financial security and prestige. This route would have been the logical choice for many in her position, especially given the expectations often placed on graduates from prestigious institutions like UDCT.

Yet, Nidhi's heart was pulling her in a different direction. The more she collaborated with S4S Technologies, the more she realized it was the perfect platform to combine her technical expertise with her desire to make positive societal impact. This path, however, meant embracing uncertainty, potentially lower pay, and moving away from the traditional career trajectory her classmates were following.

After much soul-searching, Nidhi made the bold decision to join S4S Technologies full-time upon graduating. It wasn't an easy choice – she grappled with the risks of joining a startup and the societal expectations she was defying. But her belief in the company's mission and her trust in her co-founders ultimately outweighed these concerns. Nidhi threw herself wholeheartedly into the company's early projects and product development.

Initially, S4S Technologies focused on developing and selling products for rural markets, not solely centred on agriculture. To truly understand the needs and challenges of the rural communities, the team regularly immersed themselves in rural areas of Maharashtra, particularly in the Aurangabad, Jalna, and Nagar regions. They engaged in conversations with locals and observed their daily lives. These interactions revealed a stark reality: the relentless struggle against poverty that cast a shadow over the lives of hardworking farmers.

The team discovered that the instability of agricultural income was a key driver of the poverty cycle, stemming from a complex web of factors. Unpredictable weather patterns, from droughts to floods, could destroy an entire season's crop in mere days. The fragmentation of landholdings through generations meant that most farmers were left with plots barely sufficient for their own subsistence. Moreover, post-harvest losses emerged as another significant contributor to the farmers' plight.

Despite the overwhelming odds in their lives, the farmers' resilience and hope shone through. They spoke of their dreams to educate their children, build sturdy homes, and lead dignified lives, just like their urban counterparts. However, the harsh reality was that their meagre landholdings and unstable incomes often rendered these aspirations unattainable. Families found themselves trapped in a vicious cycle of poverty, their hopes constantly shattered by their unyielding land and unpredictable skies.

Moved by the farmers' struggles, Nidhi and her team were determined to find ways to help them achieve a more stable income. While they recognized that they couldn't control the weather or land fragmentation, they identified post-harvest losses as a critical area where they could make a tangible impact.

Through extensive research and conversations with farmers, middlemen, and brokers, the team discovered that produce was being lost at every stage of the supply chain. Inadequate infrastructure, lack of preservation technology, distant markets, and improper handling, all contributed to this wastage. These post-harvest losses were not only a financial burden on stakeholders but also a significant contributor to the cycle of poverty. Every time produce was lost, it meant diminished incomes for the already struggling farmers and higher prices for consumers as the cost of the wastage was passed down the line.

Determined to break this cycle, the S4S team identified food preservation at the farmer level as a critical intervention point. They found that farmers often lacked the means to store and preserve their produce, forcing them to sell it immediately after harvest at low prices or risk spoilage. By enabling farmers to preserve their produce, the team believed they could reduce wastage, improve the farmers' bargaining power, and extend the selling window. This approach would not only provide a more stable income for the farmers but also help mitigate the boom-and-bust cycle that trapped them in poverty.

As they further explored the issue of post-harvest losses, the team became convinced that food preservation was an area where they could make a meaningful and lasting impact. They understood that by focusing on this critical aspect of the post-harvest process, they could empower farmers to take control of their livelihoods and build a more resilient future for themselves and their families. Moreover, the team's expertise in science and technology, particularly in areas such as chemical engineering and food science, made them uniquely positioned to tackle this challenge head-on.

They began exploring various methods of food preservation, including cold-chain technology and dehydration. While cold-chain solutions were effective in maintaining the quality and freshness of produce, they required a constant supply of electricity, making them impractical and cost-prohibitive for most small-scale farmers. Dehydration, on the other hand, emerged as a more viable and accessible option. By removing moisture from produce, dehydration could significantly extend its shelf life without the need for constant refrigeration.

Serendipitously, one of the co-founders, Vaibhav, had been researching dehydration as part of his university thesis and had developed a prototype for a low-cost dehydrator. Nidhi and the rest of the team joined forces with Vaibhav to refine and optimize the design for farmers, leveraging their collective knowledge of chemical engineering.

As the prototype development progressed, Nidhi took on a multifaceted role within the company. With her passion for entrepreneurship and her desire to create social impact, she focused on sales, finance, strategy, and fundraising. She learned on the job, driven by her determination to bring the dehydrator to market and make a real difference in the lives of farmers.

The path to success, however, was fraught with challenges. Once the prototype was ready, the team began selling the dehydrator to rural farmers. They soon encountered resistance and scepticism from the very people they aimed to help. Many farmers were set in their ways, relying on traditional methods like sun-drying, despite its inefficiency and the produce's vulnerability to contamination in this process. Changing entrenched habits and convincing farmers to adopt new technology proved to be an uphill battle.

Undeterred, the S4S team shifted their approach. Instead of simply pitching the product, they focused on educating the farmers about the benefits of the dehydrator. They conducted demonstrations, organized cooking competitions featuring dehydrated products, and leveraged the influence of local opinion leaders to spread awareness.

The team showcased the superior quality, nutritional value, and taste of dehydrated products compared with sun-dried ones. They emphasized how the dehydrator could help farmers preserve their produce more efficiently, reduce wastage, and potentially command better prices in the market.

Through tireless efforts and a humble approach, Nidhi and her colleagues gradually gained the trust of the farmers in the Aurangabad, Jalna, and Nagar regions of Maharashtra. Working across 400 villages, including Narla, Khamkheda, and Shelgaon, they listened to feedback, made continuous improvements to the product, and provided ongoing support and training. As more farmers adopted the dehydrator, word spread about its transformative impact on their lives.

Stories emerged of farmers who, for the first time, could save their excess harvest from spoilage. The dehydrator allowed them to preserve a variety of crops, including onions, ginger, tomatoes, garlic, and cabbage, extending the shelf life of their produce from mere days to several months. This new-found ability to store their crops gave farmers the flexibility to sell when demand was high rather than being forced to accept low prices during peak harvest season.

Moreover, the dehydrated products opened up new market opportunities. The concentrated flavours and extended shelf life attracted buyers from farther away, willing to pay a premium for quality and convenience. Farmers who had previously struggled to make ends meet found themselves with a more stable and profitable income stream.

These success stories fuelled demand for the dehydrator, and within a short period, S4S Technologies had sold over 3,000 units across 15 geographical locations. The impact of their work did not go unnoticed. The company began to receive recognition and accolades, including the prestigious Dell Social Innovation Award. This recognition provided the team with exposure and opened up new opportunities for growth and collaboration. For Nidhi, the award was more than just a validation of their hard work; it was a testament to the power of entrepreneurship in creating lasting social change.

Seeing the transformative effect of their dehydrator on the lives of farmers, Nidhi felt a deep sense of fulfilment. She realized that the true measure of their success lay not in the number of units sold, but in the positive impact they had on the communities they served. This realization strengthened her resolve to continue working towards empowering farmers and improving their livelihoods.

Inspired by the positive response to their dehydrator and its impact, Nidhi and her team made a pivotal decision. They chose to focus all their efforts on developing agricultural products and post-harvesting solutions, rather than spreading themselves thin across multiple innovations. This strategic shift marked S4S Technologies' evolution into a dedicated agri-tech company, with a clear mission to transform the lives of farmers through innovative technology solutions.

As part of this renewed focus, Nidhi, who was also in charge of sales, continued to visit villages and spend time with farmers. During these interactions, she came across another crucial insight that challenged their initial assumptions.

While some farmers had successfully utilized the dehydrator to preserve their produce and access new markets, Nidhi observed a stark contrast between large landholders and small farmholders. The former, with their extensive resources and networks, were able to find markets for their dehydrated produce and generate significant income. In contrast, small farm-holders often struggled to fully utilize the technology and secure fair prices for their crops, despite having access to the dehydration equipment.

This disparity was an eye-opener for the S4S team. They had initially believed that preventing wastage and preserving produce would automatically lead to stable income and help farmers escape poverty. However, they now realized that technology alone was not the ultimate solution to the farmers' income problems.

The team recognized that providing farmers with technology was just one piece of the puzzle. To truly make a difference, they needed to help farmers find markets for their dehydrated produce and provide them with the necessary support to navigate the supply chain effectively.

Building on these insights, the S4S Technologies team embarked on a new phase of their journey. Initially, they attempted to assist farmers by selling their diverse range of dehydrated produce to various industry players. Nidhi and her colleagues pitched everything from dehydrated spinach to chiku chips, hoping to find buyers for the farmers' products. However, they quickly realized that this scattered approach was unsustainable and ineffective. The sheer variety of produce and inconsistent quantities made it challenging to provide a reliable supply to potential corporate partners, who required predictability and consistent quality.

A turning point came when a food and beverage company proposed a transformative suggestion: S4S Technologies become an aggregator. This idea would completely reshape their business model and approach to solving farmers' challenges.

The concept of becoming an aggregator meant evolving from a mere technology provider to becoming a central hub in the agricultural value chain. As an aggregator, they would collect produce from small landholding farmers, ensure quality control, and supply it directly to corporate clients. This pivot addressed multiple challenges simultaneously:

1. It provided small farmers with access to large corporate buyers they couldn't reach individually
2. It ensured consistent quality across all products
3. It managed supply fluctuations effectively
4. It offered end-to-end support to the farmers, from dehydration technology to market linkages

Recognizing the immense potential of this pivot, Nidhi and her team embraced the aggregator role wholeheartedly. They implemented a top-down approach, first understanding client needs and then working backwards to organize their farmer network accordingly. This strategic shift allowed S4S to build stronger relationships with both farmers and corporate clients.

As they delved deeper into their new role, the team encountered myriad challenges that tested their resolve and adaptability. The transition from being a technology supplier to a comprehensive solutions provider was not without its hurdles. They had to navigate the complexities of working with numerous small landholding farmers, each with unique circumstances and constraints.

To effectively aggregate produce and ensure consistent quality, S4S Technologies invested heavily in building trust and rapport with the farming communities they served. This required spending extensive time in rural areas, engaging with farmers, and understanding their day-to-day struggles and aspirations.

It was during these immersive experiences that Nidhi and her team stumbled upon a pressing issue that had long been overlooked: the plight of women farmers. In many of the villages they visited, men were migrating to the cities in search of better opportunities, leaving the women to shoulder the burden of both household responsibilities and agricultural labour.

The challenges faced by these women farmers resonated deeply with Nidhi. As a woman herself, she empathized with their struggles, from the limited livelihood opportunities available to them to the gruelling working conditions they endured. These women had to wake up early, complete their household chores, prepare their children for the day, and then travel to work as daily-wage labourers, often in extreme heat or harsh weather conditions. After a long day of physical labour, they would return home late at night, only to start the cycle all over again, balancing household work, childcare, and elder-care responsibilities.

Moreover, the work available to these women was often inconsistent and unpredictable. They might find employment at a construction site one week, on a farm the next, and on another farm growing a different crop the following week. This lack of consistency meant that there were days when they had no work and, consequently, no income, leaving them with little control over their earnings and dependent on the availability of daily-wage jobs.

Confronted with this stark reality, Nidhi and her team found themselves at a critical juncture. They realized that their mission to transform the agricultural value chain would be incomplete without addressing the specific needs and aspirations of this often-overlooked demographic. The team decided to focus their efforts on empowering women farmers, recognizing that this approach had the potential to create a ripple effect of positive change, not just in individual lives but also in families and communities. This pivotal decision to focus on women farmers marked a bold and transformative move for S4S Technologies.

However, this new direction brought its own set of challenges. Many of the women they aimed to help lacked credit history and had no prior experience with formal financial services. Their limited savings were often consumed by household expenses and family needs, leaving them without capital to invest in food processing equipment that could potentially increase their income and improve their livelihoods.

Nidhi and her team understood that to truly empower these women farmers, they needed to provide not just the technology but also the financial support to acquire it. They approached various banks and financial institutions, hoping to secure loans for the women entrepreneurs. However, the banks were hesitant to lend money without proof of the women's creditworthiness or the profitability of their ventures.

This posed a significant challenge to the scalability of the S4S Technologies model. Without access to financing, there was no way for the women farmers to purchase the equipment needed to start their food processing businesses.

Moreover, as a startup, S4S Technologies itself had a limited balance sheet, making it difficult to convince bankers to fund the women entrepreneurs based solely on their association with the company.

Nidhi and her team explored alternative avenues for funding. They approached various grant-giving agencies, both government and private, and made a compelling case for the transformative potential of their work with women farmers. They highlighted how empowering these women with the tools and knowledge to start their own food processing businesses could not only improve their economic status but also have a ripple effect on their families and communities.

Their perseverance paid off, and they managed to secure grants to support a pilot project involving 200 women farmers. This crucial milestone allowed S4S Technologies to demonstrate the viability and impact of their model on a smaller scale before seeking larger investments.

The team threw themselves into the pilot project with enthusiasm and dedication. They provided the selected women farmers with comprehensive training on food processing techniques, including how to handle and clean the produce, the standard operating procedures for creating the final products, and the optimal times for dehydration and cutting. They equipped the women with the necessary processing equipment and worked closely with them to ensure they had the skills and confidence to use it effectively.

Throughout the pilot project, S4S Technologies provided ongoing support and guidance to the women farmers. They helped them navigate the challenges of starting and running their own businesses, from managing inventory and quality control to understanding profit-and-loss statements and maintaining proper documentation.

The results of the pilot project were remarkable. The 200 women farmers who participated successfully processed their produce using the equipment provided by S4S Technologies. They created high-quality dehydrated fruits, vegetables, and other value-added products that S4S Technologies was then able to sell to various food and beverage companies.

The income generated from these sales flowed directly to the women farmers, providing them with a stable and reliable source of revenue. The banks, upon seeing the success of the pilot and the consistent cash flows it generated, were finally convinced of the creditworthiness of the women entrepreneurs. They agreed to provide loans to help more women farmers acquire the necessary equipment and scale up their businesses.

The success of the pilot project had a profound impact beyond just financial gains. Many of the women reported a new-found sense of confidence and pride in their ability to contribute to their household income and make decisions about their own livelihoods. They felt empowered by the knowledge and skills they had gained and were excited about the prospect of growing their businesses and creating a better future for themselves and their families.

As S4S Technologies expanded its work with women farmers, Nidhi and her team developed a comprehensive approach to support them. They assisted the women in accessing government schemes, subsidies, and training programmes designed to help women entrepreneurs. They also worked with banks to identify loan products with favourable terms tailored to the women's unique needs.

The training programmes offered by S4S Technologies went beyond technical skills, focusing on entrepreneurial capabilities, financial literacy, and leadership. Guidance on financial management helped the women save, invest, and access banking services to grow their businesses.

As the women farmers' businesses grew and flourished, S4S Technologies continued to evolve its own offerings to meet their changing needs. They expanded beyond dehydration equipment to include a range of other food processing technologies, such as grain millers and milk evaporators. This allowed the women to diversify their product offerings and tap into new market opportunities.

S4S Technologies also deepened its partnerships with various stakeholders in the agricultural value chain. They worked closely with end customers, such as food and beverage companies, to understand their specific requirements and preferences. They then collaborated with the women farmers to co-develop products that met these specifications, ensuring a ready market for their goods.

To reach more women and scale their impact, S4S Technologies leveraged various channels and partnerships, working with farmers' collectives, self-help groups, NGOs, and government agencies. They developed a mobile app to connect with more women farmers and streamline the onboarding process. Nidhi and her team conducted thorough due diligence to ensure the women met the basic criteria and assessed their motivation and ability to work collaboratively.

Many women chose to work together in groups, pooling resources, sharing the workload, and supporting each other. Nidhi and her team provided them with the necessary training and support to develop effective communication and decision-making processes, setting them up for success as they embarked on their entrepreneurial journey with S4S Technologies.

Among the many inspiring stories that emerged from S4S Technologies' work with women farmers, one stood out: the heart-wrenching tale of a widowed mother of two children. Life had dealt her a cruel hand, leaving her to shoulder the responsibility of caring for her young daughter and a son with special needs who required constant attention. The weight of her circumstances was further compounded by the social stigma and isolation she faced living in her late husband's village.

Desperate to provide for her family, the mother took on the gruelling role of a daily-wage labourer in the neighbouring villages. Her long hours away from home meant that her daughter was forced to drop out of school to care for her brother, sacrificing her own future in the process. The family's plight seemed hopeless, with no end in sight to the cycle of poverty and hardship they were trapped in.

The partnership with S4S Technologies marked a turning point in this widowed mother's life. The impact was transformative. With the income she earned from processing and selling her products, she was able to provide for her family's needs and even save for her children's future. No longer did she have to travel long distances for work, allowing her to be present for her son and give him the care and attention he so desperately needed.

The nature of the work itself was also a revelation. Unlike the backbreaking labour she was accustomed to, food processing using S4S Technologies' equipment was less physically demanding. The machines took on the bulk of the work, enabling the mother to focus on quality control and other crucial aspects of running the business. For the first time in her life, she experienced a sense of control over her time and energy, a feeling that had been alien to her for so long.

Perhaps the most heartening outcome was that her daughter was able to re-enrol in school and continue her education. Freed from the burden of caring for her brother at the expense of her own future, the girl began to dream of a career in law enforcement – a path that would have been inconceivable without the opportunities created by her mother's work.

Stories like this served as a powerful testament to the transformative potential of entrepreneurship and technology in uplifting communities and changing lives. They fuelled Nidhi's belief in the work she was doing and reinforced her commitment to keeping the needs and aspirations of women farmers at the heart of S4S Technologies' mission.

As S4S Technologies continued to empower women farmers and transform lives, Nidhi found herself reflecting on the challenges she had faced as a young woman entrepreneur in the male-dominated world of agriculture. The stories of the women she worked with served as a powerful reminder of the incredible potential that could be unleashed when women were given the tools and opportunities to succeed.

However, Nidhi knew all too well that the path to success for women in agriculture was often riddled with obstacles, many of which stemmed from deep-rooted gender biases and stereotypes. As a young woman entrepreneur herself, she had experienced first-hand the scepticism and dismissal that came with being a female leader in a traditionally male-dominated industry.

Nidhi vividly recalled numerous instances when her capabilities and achievements were questioned simply because of her gender. Despite the success of S4S Technologies and the tangible impact it was having on the lives of women farmers, she often found herself on the receiving end of patronizing remarks and assumptions that she was merely running her father's or husband's business. The notion that a woman could be the driving force behind a thriving agri-tech startup seemed to be a foreign concept to many. Even when seeking funding from banks, Nidhi was subjected to questions about who the 'real' promoters or founders of the company were, as if it were inconceivable that a woman could be at the helm.

These experiences served as a poignant reminder of the pervasive gender bias and systemic inequalities that women in agriculture had to navigate daily. From being denied access to land and resources to being excluded from decision-making processes, women farmers faced numerous barriers that hindered their ability to thrive and reach their full potential.

Determined to break down these gender stereotypes and create a more inclusive and equitable environment for women, Nidhi made it her mission to challenge the status quo and advocate for greater representation and support for women in agriculture. Through her work with S4S Technologies, she sought to create a platform that not only equipped women farmers with the tools and knowledge they needed to succeed but also amplified their voices and celebrated their achievements. By showcasing the stories of women who had transformed their lives and communities through entrepreneurship, Nidhi hoped to inspire others and shift perceptions about what women in agriculture were capable of.

However, even as Nidhi and her team were making strides in women's empowerment, they found themselves grappling with a significant setback in one of their projects. Several years earlier, alongside their core focus on selling dehydrator machines and establishing market linkages for farmers, Nidhi had spearheaded the development of a direct-to-consumer (D2C) brand. The vision behind this initiative was to create a line of healthy snacks, such as beetroot, carrot, and spinach chips, made simply from dehydrated vegetables without any additives or preservatives.

Nidhi saw the D2C brand as an opportunity to tackle the widespread problem of poor dietary habits and nutrient deficiencies by providing a convenient and tasty way for people, especially children, to incorporate more fruits and vegetables into their diets. Despite the noble intentions and tireless efforts invested in this project, the D2C brand struggled to gain traction in the competitive snack market.

Over the course of four years, Nidhi and her team experimented with various marketing strategies to promote their healthy snacks. From targeting premium segments to competing head-to-head with established brands like Lay's and Haldiram's in mom-and-pop stores, they left no stone unturned. They even tried unconventional approaches such as door-to-door sales and setting up stalls at exhibitions to raise awareness about their products. Despite the high quality and nutritional value of their offerings, S4S Technologies faced an uphill battle in convincing retailers to allocate precious shelf space to their snacks, as the inventory turnover was significantly slower than for the popular, deep-pocketed brands.

The COVID-19 pandemic dealt a crushing blow to S4S Technologies, forcing them to shut down their D2C brand. It left Nidhi grappling with self-doubt and questioning whether they had missed something crucial in their approach. This setback coincided with a slowdown in their dehydrator equipment sales, as the team was in the midst of pivoting to focus on women farmers.

To make matters worse, securing funding proved to be another significant hurdle for the agri-tech startup. As they focused on hardware solutions, most investors struggled to categorize S4S Technologies, unsure whether to classify it as an equipment manufacturer, a food processing company, an agricultural venture, or a livelihood-focused enterprise, making it challenging for the venture to attract capital investment.

It was during this critical juncture that the value of mentors became apparent to them, particularly of Hemendra Mathur, whom Nidhi affectionately refers to as the 'Bhishma Pitamaha' of agriculture. Mathur's guidance and extensive network were invaluable in helping S4S Technologies navigate the funding landscape, connecting the team with the right investors who understood and aligned with their vision and mission.

Through Mathur's connections, S4S Technologies found an investor who stood out for their comprehensive understanding of the challenges involved in development work in India. With the backing of this investor and the continued support of mentors like Hemendra Mathur, S4S Technologies secured seed funding of ₹4.8 crores. This marked a turning point in their journey as they focused their efforts on creating a comprehensive ecosystem to support and empower women entrepreneurs in rural India.

As S4S Technologies continued to grow and evolve, Nidhi remained steadfast in her vision of building a sustainable business that could create a lasting impact while ensuring financial viability. The company operated on a robust revenue model that supported its mission to empower farmers, particularly women.

S4S Technologies' primary income stream came from its role as an aggregator in the agricultural value chain. By purchasing dehydrated produce from its network of women farmers at fair prices and selling these high-quality products to food and beverage companies, the company generated steady revenue. This model provided a reliable market for farmers while allowing S4S to scale its operations and impact. The company also sold food processing equipment, including dehydrators and other technologies, to farmers and small-scale entrepreneurs.

This approach allowed S4S Technologies to balance profitability with social impact. As a result, the company could attract investments, scale its operations, and create a sustainable model for rural development.

Navigating the challenges and setbacks in her entrepreneurial journey, Nidhi found herself drawing parallels between the world of sports and the mindset needed to succeed in a startup. She admired the spirit and tenacity of incredible athletes like Michael Jordan, Muhammad Ali, Serena Williams, and Roger Federer, who constantly pushed themselves to improve and reach new heights. Nidhi realized that entrepreneurs must cultivate a similar learning mindset and a relentless drive to keep getting better, even in the face of adversity.

Looking back, Nidhi recognizes that her professional journey with S4S Technologies has also been a deeply personal one. Through her work, she discovered her true purpose and what mattered most to her: making the world a little bit better through small, concerted efforts. As S4S Technologies moves forward, Nidhi is determined to build a company that functions like a sports team, with everyone aligned around a common goal and driven by the impact they can create for the people they serve. She wants her team to be motivated not by their allegiance to her or the company but by their genuine care for the farmers whose lives they are working to improve.

To this end, Nidhi and her co-founders are committed to hiring the right people and partnering with those who share their values and vision. They believe that building a team of individuals who are passionate about creating positive change is essential to achieving their mission.

Most importantly, Nidhi has come to understand that true success lies not in personal achievements or financial gains, but in the impact one can create in the world. She measures her own success by the lives she has touched and the positive change she has been able to bring about through her work.

As she reflects on the future of S4S Technologies, Nidhi is excited about the possibilities that lie ahead. She envisions a world where every farmer, regardless of their gender or socioeconomic status, has access to the tools, knowledge, and resources they need to thrive. She sees a future where rural communities are empowered to build a better life for themselves.

# 8

# The Unlikely Champion of Innovation: Shridhar Mehta, Prompt Dairy Tech

*E*VERY DAY, MILLIONS OF *Indians stir milk into their chai, add it to their children's cereal, or pour themselves a glass. We recognize the farmers who care for the cows, the cows themselves, and the brand names on the milk packets. But do we ever wonder about the unseen heroes working tirelessly behind the scenes to ensure that the milk reaching our table is safe, pure, and of the highest quality? For over three decades, one man and his team have been at the forefront of revolutionizing India's dairy industry, innovating relentlessly to bring us healthier, fresher, and more nutritious milk. This unsung hero is Shridhar Mehta, a visionary entrepreneur whose story of perseverance and innovation has touched the lives of millions, though few know his name. This is his remarkable journey.*

In the bustling city of Ahmedabad in Gujarat, a young Shridhar Mehta grew up with the weight of expectations on his shoulders. Born in Baroda but raised in Ahmedabad, he was the son of a man who dared to dream beyond the confines of a government job. In 1978, Shridhar's father took a leap of faith, leaving the security of his position to start a business selling power tools. This bold move would shape the family's future and unknowingly set the stage for Shridhar's own entrepreneurial journey.

As the family's power tool business flourished, his father harboured hopes that his children would one day take over and expand the family enterprise. But young Shridhar had different ideas brewing in his mind, even if he couldn't quite articulate them yet.

In school, the future entrepreneur was far from a star pupil. His academic performance was average at best, often drawing discouraging remarks from his teachers. 'You won't amount to much,' they would say, their words stinging but never truly penetrating his resilient spirit. What these educators failed to recognize was that Shridhar's apparent disinterest in academics wasn't due to a lack of capability but rather an indication of a mind yearning for a different kind of knowledge.

The 1980s brought winds of change to India, and with them came the first whispers of a technological revolution. As Shridhar entered college to study commerce in 1985, he encountered something that would change the trajectory of his life: computers. At a time when India's economy was still largely closed and the digital age was in its infancy, he found himself irresistibly drawn to these mysterious machines.

In 1985, India was a far cry from the tech hub it would eventually become. Computers were rare, expensive, and understood by few, while software was an alien concept to most. Those intrigued by these mysterious machines were often met with a blend of curiosity and scepticism. Yet for Shridhar, it was love at first sight, igniting in him a passion that would drive him to become a pioneer in the tech industry.

With a determination that belied his lacklustre academic record, the young student threw himself into learning everything he could about computers. He started with COBOL and BASIC, two programming languages that were the building blocks of early computing. The learning process was arduous. There was no internet to turn to for quick answers, no YouTube tutorials to guide him through tricky concepts. Books were scarce, and those who understood computer technology were even scarcer.

Shridhar's journey into the world of computing was a solitary one, filled with trial and error. He would spend hours poring over manuals, experimenting with code, and trying to decipher error messages. Each small victory – a programme that finally ran without crashing, a function that worked as intended – was celebrated as a major triumph.

As his knowledge grew, so did his passion for sharing it. In a bold move that foreshadowed his future as an entrepreneur, Shridhar decided to start a computer-teaching organization while still in college. He named it Zalak Computer Centre, marking his first foray into the world of business.

Running Zalak was no small feat for a college student. The young entrepreneur had to balance his studies (which, admittedly, took a back seat) with the demands of teaching and managing a fledgling business. He faced the challenge of convincing sceptical parents that computers were the future and that their children would benefit from learning these skills. Resources were scarce, requiring ingenuity to acquire the necessary equipment and create a curriculum that would make sense to students who had never seen a computer before. Despite these obstacles, Shridhar thrived on these challenges. For the first time in his life, he felt truly engaged and excited about learning and teaching.

After graduation, duty called, and the young graduate joined his father's power tool business. But even here, his passion for technology found an outlet. Recognizing the inefficiencies in the business' operations, he took it upon himself to write software for inventory control and sales tracking. This system not only kept accurate stock of power tools but also analyzed sales patterns, helping to optimize inventory levels and predict future demand.

It was his first taste of how technology could transform traditional industries, a theme that would define his future endeavours.

As the 1990s dawned, Shridhar's father, ever the forward-thinking businessman, decided it was time to diversify the family business interests. He tasked Shridhar and his brother, a petrochemical engineer, with exploring new business opportunities. The brothers, eager to make their mark, dove into research, attending trade shows, meeting with industry experts, and analyzing market trends.

Their search led them to the then-booming electronics sector. Despite neither of them having a background in electronics, the Mehta brothers were undeterred. They saw potential in it and were willing to learn.

The learning curve was indeed steep. The siblings immersed themselves in the world of electronics, meeting with experts, poring over technical documents, and trying to understand the intricacies of product development. Their persistence paid off when they identified a promising product: electronic weighing scales.

The decision to focus on weighing scales was not arbitrary. They saw a market ripe for disruption, where traditional mechanical scales could be replaced by more accurate and efficient electronic versions. With characteristic thoroughness, they hired technical experts and prominent consultants, aiming to create a product that would stand out in the market.

In 1992, after months of development and countless setbacks, Shridhar and his team launched their first electronic weighing scale. It was a proud moment, made even sweeter by the fact that they were the first company in Gujarat to receive a licence for manufacturing electronic weighing scales. The product line was diverse, encompassing scales for industrial applications as well as precision instruments for specific sectors like jewellery.

As the weighing scale business gained momentum, Shridhar found himself managing an increasingly complex portfolio. He continued to oversee the Zalak Computer Centre while simultaneously steering the new electronics business and contributing to his father's power tool company. This period marked intense growth, both for the various business ventures of the family and for Shridhar's development as an entrepreneur.

Amidst the success of the weighing scale venture, there arose an unexpected opportunity that would set Shridhar on the path of his true calling. Among the customers for their scales were buyers from the dairy industry. These clients approached him with a specific request: could he develop a scale that would measure milk volume?

This inquiry prompted Shridhar to delve deeper into the problem. The milk collection process was a daily struggle for millions of farmers across the country. They would bring their milk to the village collection centres, where it was measured using simple volume cups – a method highly vulnerable to inaccuracies and manipulation. Farmers often left feeling short-changed, while collection centres grappled with inconsistent measurements.

Recognizing an opportunity to make a difference, Shridhar understood that the density of milk varied with its fat content, making weight-to-volume conversion more accurate than direct volume measurement. Armed with this knowledge, he and his team set out to create a groundbreaking solution that would bring transparency and fairness to the process.

The result was an innovative weighing scale that could convert kilograms to litres, specifically calibrated for milk measurement. The digital display showed clear, indisputable measurements, bringing unprecedented transparency to the process. Collection centres benefitted from the resultant increase in accuracy and efficiency, while farmers could now see and trust the exact measure of their production.

The impact was immediate and profound. As word spread, more and more village collection centres adopted the new scales. The success of this innovation ignited something in Shridhar. He found himself increasingly drawn into the world of dairy, fascinated by the complexities of the supply chain and the challenges faced by farmers.

Driven by this new-found interest, Shridhar began making frequent trips to the villages and milk collection centres. He was determined to understand every aspect of the dairy industry, from the smallest farmer to the largest processing plant. What he witnessed on these visits both shocked and inspired him.

In these rural communities, Shridhar encountered a world far removed from the bustling city life he knew. He saw farmers who couldn't afford to keep the milk from their cows for their own children and were forced to sell every drop to make ends meet. In one village, he met a family where the mother diluted the little milk they kept with water, stretching it to feed her four children. The sight of these malnourished children, living in the heart of a milk-producing region, left an indelible mark on Shridhar's conscience.

The root of this paradox lay in the inefficient and often exploitative milk collection process. Farmers, the backbone of India's dairy industry, were trapped in a system fraught with manual interventions and opaque record-keeping. Their milk deliveries were recorded in small, easily manipulated notebooks, leaving them vulnerable to errors and fraud. Without access to their historical data, farmers had no way to verify their payments or challenge discrepancies.

This lack of transparency had far-reaching consequences. Farmers struggled to manage their daily expenses, let alone invest in their farms. Their financial insecurity bred stress and frustration, rippling out to affect entire families and communities. Payments, calculated every ten days, were often riddled with errors, leaving farmers feeling cheated and demoralized. Shridhar saw families teetering on the edge of financial ruin, their livelihoods hanging by a thread as a result of these systemic inconsistencies.

The more he observed, the more Shridhar realized that the weighing scale was just the tip of the iceberg. The entire milk collection process needed a complete overhaul. He envisioned a system that could automate record-keeping, ensure transparency, and provide farmers with real-time access to their data. Such a system could not only prevent exploitation but also empower farmers with the information they needed to make informed decisions about their businesses.

These thoughts ignited a fire within Shridhar. Despite the success of his electronic weighing machine business, he felt a growing urge to do something more impactful, something that could transform the lives of these farmers. The seed of an idea began to take root in his mind – what if he could apply his knowledge of technology to automate and streamline the entire milk collection process?

With the same determination that had driven him to learn computing a decade earlier, Shridhar threw himself into researching and developing a milk collection system for village collection points. However, the path ahead was far from easy, especially in rural India in the mid-1990s.

Abandoning the comforts of city life, Shridhar immersed himself in the world of dairy farmers in the villages. He slept on hard floors in community centres or in the homes of generous villagers, bathing in a different house each day. This stark shift from his usual lifestyle served as a constant reminder of his mission's importance.

As days turned into weeks and weeks into months, Shridhar's frequent absences and seemingly endless village stays began to raise eyebrows. Family members and his father's business associates voiced their concerns. 'Why is he wasting time in the villages when the weighing machine business is booming?' they wondered. They saw only uncertainty where Shridhar envisioned transformation.

The product development process proved to be a gruelling test of perseverance. In an age before CAD software and 3D printing, each iteration demanded painstaking manual crafting and testing. Shridhar set up a makeshift workstation in a village, assembling a computer to write automation software. But the rural infrastructure fought him at every turn. Power cuts were frequent and merciless, often leaving him working late into the night by lamplight. Hard drives fell victim to voltage spikes, forcing long trips to the nearest town for replacements. Even UPS systems succumbed to the erratic power supply.

Yet, for every setback, there was a moment of kindness that fuelled Shridhar's resolve. The villagers, despite their own hardships, opened their homes to him and shared their meals with him. Their generosity strengthened his determination to create a system that would significantly improve their lives.

As the months wore on, financial pressures mounted. The new venture consumed resources without generating income, and Shridhar's future looked increasingly uncertain. But the faces of the farmers he'd meet, their stories of struggle and hope, drove him forward.

Finally, after countless iterations and sleepless nights, Shridhar's persistence paid off. In 1995, he unveiled the prototype of his automated milk collection system. This dot-based system seamlessly integrated electronic weighing machines with computers, automatically recording and storing data that had previously been painstakingly logged by hand.

The system's impact was immediate but not universally embraced. While many farmers approached milk collection with new-found confidence, knowing every drop would be accurately accounted for, some collection centres resisted. These centres, entrenched in their old ways and sometimes profiting from the opacity of manual systems, were reluctant to adopt the new technology. Shridhar found himself facing a new challenge: convincing these holdouts of the long-term benefits of transparency and accuracy.

He redoubled his efforts. He organized demonstrations in the villages to show farmers and collection centre operators how the system worked. He patiently addressed concerns and highlighted the long-term benefits for all parties involved. Gradually, more milk collection centres began to adopt the technology, seeing how it improved trust and efficiency in the milk collection process. Despite dairy technology accounting for only 10 per cent of his business at the time, with electronic weighing machines still the mainstay, Shridhar's unwavering commitment to improving farmers' lives drove him to focus on this crucial segment.

As Shridhar continued his work in the villages, he uncovered another pain point: milk fat-content measurement. The existing process, even with some electronic devices, still relied heavily on manual intervention. This led to inconsistencies and bred distrust among farmers, who were often sceptical about the reported quality of their milk.

Never one to shy away from a challenge, Shridhar set his sights on solving this problem too. In 1999, he began developing the 'Fat'omatic' – a fully automatic fat testing machine. This groundbreaking device could draw milk samples, measure their fat content, and integrate the results with a computer system, all without human intervention.

The Fat'omatic quickly gained traction, impressing both farmers and collection centres with its accuracy and efficiency. However, Shridhar soon realized that many smaller collection centres in villages couldn't afford such sophisticated systems. So, he adapted his innovation, creating a microprocessor-based version that significantly reduced costs while maintaining reliability. This move exemplified Shridhar's commitment to inclusive innovation, ensuring his solutions could benefit farmers across all economic strata.

Shridhar's innovative spirit, however, was far from satisfied. As he continued to immerse himself in the dairy industry, he identified yet another crucial challenge: the need for rapid milk-density assessment. Traditionally, farmers had to wait for separate tests for fat content and density, often leading to long queues at collection centres and delays in payment. Shridhar envisioned a solution that would revolutionize this process, benefitting both farmers and the industry at large.

While the Fat'omatic had automated fat-content measurement, milk density was still being measured using a lactometer, a process prone to manual errors and inconsistencies. Shridhar recognized that combining these two crucial measurements into a single, automated process could revolutionize milk-quality assessment.

In 2006, after months of meticulous research and development, Shridhar unveiled India's first indigenous ultrasonic milk analyzer. This groundbreaking machine was a game changer, capable of measuring fat content, solid non-fat content, added water, and density – all within an astonishing forty-five seconds. By automating both fat and density measurements, the analyzer eliminated the need for manual lactometer readings, significantly reducing the potential for human error and manipulation.

For farmers, this meant more than just quicker testing. It ensured a fair, transparent assessment of the quality of their milk, free from the vagaries of manual intervention. Gone were the long waits and uncertainties associated with separate fat and density tests. Now, farmers could receive accurate results and fair payments promptly, boosting their confidence in the system and enhancing their financial security.

Collection centres, too, reaped the benefits of this innovation. The all-in-one analyzer streamlined operations, reduced labour costs, and minimized disputes over milk quality. The automated data recording directly into computer systems further enhanced transparency and traceability throughout the supply chain.

As word of this revolutionary dairy technology spread, the demand for Shridhar's company Prompt Dairy Tech's products surged beyond Gujarat's borders. Shridhar saw an opportunity to make a broader impact and began strategically expanding his company's reach across India. He established a network of distributors and service centres, aiming to ensure his technologies could be effectively deployed and maintained across different regions.

However, this rapid expansion brought its own set of challenges. While the distribution network initially seemed like a logical solution to meet the growing demand, it soon became apparent that many distributors, despite receiving good margins, were failing to provide adequate service to customers. In an era of limited communication channels, farmers struggling with equipment issues had no direct line to the company, which threatened to tarnish Prompt's hard-earned reputation.

Faced with this dilemma, Shridhar made a bold and counterintuitive decision. Despite the financial risks involved, he chose to bring all servicing in-house. This meant employing technicians and strategically positioning them across rural areas to address a wide range of maintenance needs, regardless of the job's scale or immediate profitability. It was a move that defied short-term financial logic but aligned perfectly with Shridhar's commitment to farmer welfare and long-term vision for the industry.

This decision marked a turning point for Prompt Dairy Tech. By prioritizing service quality and direct farmer relationships over short-term profits, Shridhar not only protected his company's reputation but also deepened its connection with the farming community. It was a testament to his belief that true innovation goes beyond creating new technologies – it's about ensuring those technologies genuinely improve lives and build trust across the entire dairy supply chain.

However, this expansion and reorganization pushed Prompt Dairy Tech to the brink of financial collapse. The company's unconventional focus on rural technology and its commitment to in-house servicing made it appear risky to traditional lenders. Banks, sceptical of the business model, refused to extend credit. Shridhar found himself turning to friends and family, borrowing at exorbitant interest rates just to keep the company afloat.

The weight of these financial obligations bore down on Shridhar relentlessly. Each month, as the 7th approached – payday for his growing workforce – and the 10th loomed – the deadline for interest payments – Shridhar felt the vice of anxiety tightening around him. He would pace up and down on sleepless nights, crunching numbers, trying to stretch every rupee to meet these critical deadlines.

The high costs of research and development, coupled with the expensive process of manufacturing precision instruments, meant that profit margins were razor-thin. Each new product iteration required significant investment, with returns often delayed by months or even years. This cycle of innovation, while crucial for staying ahead in the market, put immense strain on the company's finances.

To compound these challenges, Shridhar encountered unexpected financial discrepancies at the operational level.

At a time when cash transactions were the norm across most Indian businesses, particularly in rural areas, some employees succumbed to temptation, pocketing money from milk collection centres. The lack of robust monitoring systems, a consequence of rapid expansion, had left the company vulnerable to such leaks.

As losses mounted and debts piled up, Shridhar found himself at a crossroads. The voices of doubt grew louder, with even his family urging him to cut his losses and shut down the business. It was a dark moment, one that tested the very core of Shridhar's resolve and vision.

But in this moment of crisis, his entrepreneurial spirit shone through. Instead of capitulating, he chose to adapt. Shridhar realized that to continue his mission of helping farmers, he first needed to ensure his business's survival. With a heavy heart, he made the difficult decision to pause the relentless cycle of innovation and product development that had defined Prompt Dairy Tech. Instead, he turned his focus inward, towards consolidation and process improvement.

This period of introspection and restructuring became a transformative phase for the company. Shridhar meticulously examined every aspect of the business, from supply chain to customer service. He implemented technological solutions to plug financial leaks, streamlined operations to reduce costs, and overhauled the cash flow management system. Gradually, painstakingly, he steered the company back towards profitability.

As Prompt Dairy Tech emerged from this crucible, leaner and more resilient, Shridhar's innovative spirit reignited. In 2011, he identified a critical gap in the dairy supply chain's data management system. While individual collection centres had desktop-based data, there was no centralized system for real-time visualization at the dairy level. This fragmentation meant that crucial information was trapped at local stations, hindering efficient decision-making and management across the supply chain.

Shridhar set out to revolutionize the industry once again. In an era when cloud computing was still in its infancy, especially in rural India, he envisioned a system that would connect every milk collection centre to the cloud. With only 2G networks available in most villages, this was no small feat. Shridhar's determination knew no bounds. He collaborated with Vodafone to create a virtual private network, effectively bringing the power of cloud computing to the remotest corners of Gujarat. This groundbreaking initiative made Prompt Dairy Tech the first to develop a cloud-based system in India's dairy supply chain, setting a new standard for the industry.

The impact of this cloud-based innovation was immediate and profound. For the first time, stakeholders across the dairy supply chain had access to real-time data, transforming decision-making and operational efficiency. A leading national dairy federation took notice, initiating a pilot project to meticulously observe the system's benefits over a two-year period.

The results were staggering. Not only did the cloud-based platform improve transparency and reduce inefficiencies but it also slashed time to market and significantly reduced milk wastage. Impressed by these outcomes, a national dairy federation made a landmark decision in 2015, awarding Prompt Dairy Tech a contract to install the system in all its 18,000 milk collection centres across Gujarat.

This massive project catapulted Prompt Dairy Tech to new heights. The platform soon became the backbone of Gujarat's dairy industry, with an astounding 3.6 million farmers updating their milk deliveries twice daily. The sheer scale of this operation – 3.5 million farmers interacting with the system every morning and evening – created a data-rich ecosystem that benefitted the entire supply chain.

But Shridhar's innovative spirit was far from satisfied. His attention turned to a persistent problem plaguing the industry: milk adulteration. This issue not only posed health risks to consumers but also unfairly penalized honest farmers in cooperative societies who often bore the brunt of others' misdeeds.

Collaborating with FOSS Analytical, a global leader in analytical instruments, Shridhar and his team spent nearly a decade developing MilkoScreen. This compact marvel could measure basic milk parameters and detect various adulterants within forty-five seconds. The key innovation was the creation of a 'pure milk fingerprint' – a baseline standard against which all milk samples could be compared. This revolutionary approach allowed for rapid and accurate detection of even the subtlest forms of adulteration.

Installed in over 10,000 villages, MilkoScreen democratized high-quality testing, making it accessible even to small collection centres. The impact was transformative. Milk quality improved across the board, and farmers gained a deeper understanding of best practices. For honest farmers, it meant recognition and fair compensation for their quality produce. For consumers, it ensured safer, higher-quality milk.

As Shridhar's innovations continued to reshape the industry, he turned his attention to another critical link in the dairy supply chain: milk chilling centres. These facilities faced numerous challenges, including inefficient temperature management, lack of real-time monitoring, and difficulties in tracking milk quantity and quality during storage and transportation.

Recognizing these issues, Shridhar and his team developed an IoT system for monitoring bulk milk containers. With over 70,000 bulk milk coolers already installed across India, this system was a game changer. It provided real-time data on milk quantity, temperature, and even the compressor efficiency of the coolers. For an industry dealing with a highly perishable product, this level of monitoring was revolutionary. It minimized milk spoilage, optimized transportation, and ensured that high-quality milk reached processing plants in perfect condition.

This innovation, along with Prompt Dairy Tech's earlier developments, had a profound impact on the entire dairy supply chain. From individual farmers to massive processing plants, the company's technology was transforming operations at every level. As a result, Prompt Dairy Tech solidified its position as an industry leader, with its solutions becoming integral to the modernization of India's dairy sector.

The company's rapid growth and diversification led to a complex corporate structure. By 2017, Shridhar found himself at the helm of three or four companies, each specializing in different aspects of dairy technology. While this structure had allowed for targeted innovation and specialized expertise, it was becoming increasingly challenging to manage efficiently.

On the advice of a trusted mentor, Shridhar made the bold decision to consolidate these entities into a single, unified company. This merger, while strategically sound, presented its own set of hurdles. Aligning the visions and decision-making processes of various company directors proved to be a complex task. Once again, Shridhar's mentor provided sage advice: it was time to professionalize the management structure.

In a move that many entrepreneurs find challenging, Shridhar decided to bring in experienced professionals to run the day-to-day operations of the company. This decision marked a significant turning point for the company. By recruiting seasoned executives, including a new CEO, Shridhar was able to create a more coherent roadmap for the company's future. This restructuring allowed him to focus on his true passions: strategic development and deepening customer relationships.

The results of this organizational transformation were remarkable. Today, Prompt Dairy Tech boasts an impressive portfolio of over thirty-five products serving the entire dairy supply chain. The company's reach extends to 70,000 villages across twenty-eight states and over 350 districts in India. With a workforce exceeding 1,000 employees and projections of ₹500 crore for the coming year, Prompt Dairy Tech has emerged as a powerhouse in the Indian dairy tech industry.

Yet, for Shridhar, this success only fuelled his drive to further innovate and improve the lives of farmers. With the company now operating more efficiently, he could turn his attention to addressing challenges at the grassroots level of dairy farming. This shift in focus led to the development of new technologies aimed at improving animal health and productivity.

One such innovation was the development of smart tags for animal health monitoring and heat detection. These tags allow farmers to track the health and reproductive cycles of their cattle with unprecedented accuracy, optimizing breeding practices and reducing veterinary costs. Another groundbreaking product was a rapid pregnancy testing kit for cattle, enabling farmers to detect pregnancy in cows much earlier than traditional methods.

These innovations not only improved the efficiency of dairy farming but also directly impacted the quality of milk entering the supply chain. By enhancing animal health and breeding practices, Prompt Dairy Tech was addressing milk-quality issues at their source, complementing the existing solutions for milk collection and transportation.

As the company continued to grow, Shridhar's keen eye for identifying and solving problems in the dairy supply chain never wavered. His next focus was on addressing the high bacterial load in milk during transportation from collection centres to chilling stations – a persistent issue that compromised milk quality and limited export potential.

The impact of high bacterial load was far-reaching. Farmers risked having their milk rejected or receiving lower prices for poor-quality milk. Processors faced increased costs due to spoilage and shorter shelf life, while consumers were potentially exposed to health risks from contaminated products.

Determined to address this challenge, Shridhar and his team developed MilkoChill, an innovative product capable of rapidly cooling milk at the collection centre itself. This solar-powered device can chill 40 litres of milk in just 10 minutes, compared to the 3 hours required by traditional bulk coolers. The innovation was so significant that it won the Animal Husbandry Startup Grand Challenge 2.0 organized by the Department of Animal Husbandry and Dairying.

Today, collection centres across India have adopted this system, dramatically improving milk quality and reducing spoilage. By ensuring that milk reaches chilling centres already cooled, MilkoChill has effectively halted bacterial growth during the critical transportation window.

This innovation exemplified Prompt Dairy Tech's holistic approach to improving the dairy supply chain. While the company's primary focus remained on B2B relationships with major players like Amul and Nestlé, the ripple effect of their innovations has reached millions of farmers and consumers across India.

The innovative spirit at Prompt Dairy Tech extended far beyond cooling solutions. The company developed sophisticated calibration equipment to ensure the accuracy of milk-quality testing machines across thousands of collection centres. Alert systems were created to notify farmers and stakeholders of any anomalies in milk quality or quantity, enabling swift identification and resolution of issues.

Perhaps one of the most transformative innovations was the Farm365 app. This free app gave farmers unprecedented access to their data, from milk production records to animal health information. More than just a management tool, the app became a gateway to financial inclusion, providing farmers with verifiable production records that made securing loans easier than ever before.

The cumulative impact of these technological advancements was profound. Farmers who once managed small herds of one to two livestock were now able to expand their operations significantly. Improved breeding cycles and animal health monitoring led to increases in both milk production and quality. Moreover, these innovations fostered a more equitable system within cooperative societies, rewarding and protecting good farming practices.

While these advancements have revolutionized the dairy industry, their impact often goes unnoticed by the average consumer. Every time we purchase a packet of milk from a major brand, chances are it has passed through one of Prompt Dairy Tech's machines. This silent revolution in milk quality and safety touches the lives of millions of Indians daily, from infants to the elderly, yet few are aware of the man behind this transformation.

For over two decades, Shridhar has been working tirelessly in one of India's most challenging sectors – agriculture. Unlike many entrepreneurs who seek the limelight, he finds satisfaction in the quiet corners of rural India, developing products and spending time with the farmers he serves. His joy comes not from public recognition but from moments like meeting a farmer from a village where he installed a machine in the 1990s and hearing first-hand how that early intervention changed the farmer's life for the better.

Shridhar's journey from a young man struggling with traditional academics to the leader of a company that has revolutionized India's dairy industry embodies the transformative potential of technology when utilized with empathy and understanding. His story is one of relentless innovation driven by his deep connection to India's dairy farmers.

Today, as Prompt Dairy Tech continues to push the boundaries of what's possible in dairy technology – working to improve milk quality to meet international standards and collaborating with IITs on cutting-edge research – Shridhar Mehta's vision remains unchanged. For him, success is not measured in the crores of rupees earned or the accolades received, but in the number of farmers' lives improved and the quality of milk reaching Indian households.

As we enjoy our daily glass of milk or cup of tea, it's worth remembering the unseen innovator whose dedication ensures its quality and safety. Shridhar Mehta's legacy extends far beyond the dairy fields of India, offering inspiration to entrepreneurs and innovators worldwide seeking to make a tangible difference in their communities. It's a narrative that deserves to be told and celebrated, both within and beyond the agricultural sector, as a shining example of purposeful entrepreneurship that truly changes lives.

# 9

# The Barefoot Entrepreneur: Shashank Kumar, DeHaat

*IN THE VAST LANDSCAPE of human existence, where countless lives intertwine and diverge, the story of Shashank emerges like a river carving its path through uncharted terrain. Born into a world of scarcity and hardship, his journey is a testament to the transformative power of resilience, determination, and unwavering belief in the value of education. Just as a river's course is shaped by the rocks and valleys it encounters, Shashank's life has been moulded by the challenges and experiences of his humble beginnings in a small town in Bihar. This barefoot boy, who once navigated the rugged terrain of rural poverty, would go on to revolutionize Indian agriculture, his footsteps leaving an indelible mark on the fields he sought to transform.*

Shashank Kumar was born in 1985 in a small town called Chapra in the heart of Bihar's Saran district. His was a lower-middle-class family. His father, a grade-4 employee in the state electricity department, and his mother, a local middle-school teacher, worked hard to support their three children. The family owned a small piece of farmland, but it was barely enough to supplement their modest income.

Growing up as the youngest child, Shashank witnessed first-hand the daily struggles of rural life in India. The family's circumstances were typical of the area at that time – a life marked by simplicity, hardship, and resilience. Despite the scarcity that permeated every aspect of their existence, Shashank's parents held firm to a powerful belief: education was the key to breaking the cycle of poverty.

Life in their remote village was a constant lesson in adaptability. Even simple tasks posed significant challenges. A trip to the nearby town, a mere 40 km away, was an arduous journey involving a 5-km walk to the bus stop followed by a four-hour ride. Yet, like their neighbours, Shashank and his family accepted these hardships as an integral part of their lives.

The young boy quickly learned to make the most of the limited resources available to him. As he had only two school uniforms, he took meticulous care of his belongings. As darkness fell each evening, plunging the village into blackness, Shashank and his siblings would huddle around a flickering lantern to study. These nightly study sessions, illuminated by the soft glow of the lantern, became a symbol of their determination to learn despite the odds.

From an early age, Shashank was taught the value of responsibility and independence. At just seven, he was expected to contribute to the household chores alongside his siblings. Washing clothes, polishing shoes, organizing bookshelves, and even maintaining the precious lantern became part of his daily routine. These experiences, while demanding for a young child, laid the foundation for Shashank's understanding of the world.

The young boy's ability to navigate life's demands was tested when his elder brother cleared the competitive exams for the fifth grade at the prestigious Sainik School. The entire village buzzed with excitement, congratulating the family on his remarkable achievement. It was a moment of immense pride, as villagers recognized the significance of a boy from their humble community securing a place in such a competitive institution.

While initially sharing in the happiness and pride, Shashank later overheard a conversation between his parents.

In the quiet stillness of their small house, they discussed the daunting challenge of affording the school fees. The financial burden weighed heavily on their minds as they grappled with the realization that arranging the necessary funds would be a monumental struggle.

Listening to his parents' concerns, the boy's heart sank. He realized that beneath the celebration and joy, his family faced a harsh reality. The pursuit of a better life through education came at a cost, one his parents were willing to bear but not without immense sacrifice. This moment was etched into his young mind.

The following year, as Shashank prepared for his own entrance exam to the school, he was determined to find a way to alleviate his parents' financial burden. Knowing his family couldn't afford fees for both him and his brother, he sought an alternative path. His search led him to the Netarhat Residential School in Jharkhand, which offered a full scholarship to students excelling in its highly competitive entrance exam. This presented an opportunity to secure his education without further straining his family's finances.

In 1996, at just eleven years of age, he sat for this challenging exam, competing with 12,000 students for a mere seventy seats. Despite the odds, his determination and hard work paid off. Coming from a remote village where he studied under the light of the lantern and balanced academics with household chores, the young boy defied expectations and cleared the exam.

Years later, this early experience of resilience and frugality served Shashank well as he embarked on his entrepreneurial journey. After completing his education at IIT and working for three years in a reputable company, he carried these valuable lessons into his new venture. For eight years, he bootstrapped his startup, working tirelessly with farmers at the grassroots level.

Resources were scarce in the early days of the venture, mirroring the conditions of Shashank's youth. He and his crew lived frugally, staying in villages with the farmers they served. For nearly three years, he slept on school benches every night, a situation that might have daunted others but felt familiar to him.

The numerous rejections and setbacks faced during this period were all part of a journey Shashank had been unknowingly preparing for since childhood. Growing up, hearing 'no' was commonplace, and he had learned not to let it discourage him. His upbringing had instilled in him a resilience that allowed him to persevere without needing external motivation. When people pointed out that his experiences were typical entrepreneurial struggles, Shashank would laugh it off. To him, it was routine, a normal part of life that he had been accustomed to since his early years. The first seven or eight years of his life had laid the foundation for the resilience and growth mindset that would carry him through the initial eight years of his entrepreneurial journey.

This adaptability was further shaped by his time at Netarhat Residential School from 1997 to 2003. The boarding school environment exposed him to a diverse group of students from various backgrounds and economic conditions, providing him with a wealth of knowledge and perspective that differed significantly from what his village life would have given him.

From sixth to twelfth grade, Shashank thrived at Netarhat, excelling academically while immersing himself in extracurricular activities. He played hockey, participated in dramatics, and spent countless hours in the library, ensuring his own well-rounded development. The school's gurukul concept, emphasizing self-reliance and responsibility, reinforced the values of hard work, humility, and equality that he had begun to learn even as a child.

It was during his time at Netarhat that he first learned about the IITs and the promise they held for a bright future for those who were admitted into them. The school's ecosystem was conducive to preparing for these prestigious institutions, with many students aspiring to secure a spot in them. Shashank discovered that success at an IIT could lead to government stipends that cover fees, a well-paying job, financial stability, a prospect that greatly motivated him.

Driven by this new-found goal, he dedicated himself to his studies. However, his first attempt at cracking the JEE to secure admission to an IIT was unsuccessful. This setback, while disappointing, didn't deter him. As a backup, he had also applied to the Cochin University of Science and Technology (CUSAT) and secured a good rank in their entrance exam. This achievement brought him some relief, as CUSAT was also a reputable institution.

Coincidentally, Shashank's elder brother had also cleared the CUSAT exam after taking a year off to prepare for it. The prospect of studying together for the next four years thrilled both siblings, an opportunity they rarely had as they had attended different boarding schools.

Fate, however, had other plans. Financial constraints forced the family to make a difficult choice. They could only afford to send one brother to university. This realization came as a shock to Shashank, but his resilience, honed through years of challenges, allowed him to adapt quickly. With unwavering resolve, he decided to give the IIT entrance exam another shot, taking a year to prepare while his brother pursued his education at CUSAT.

During this pivotal year, Shashank channelled all his energies into preparing for the JEE. His dedication paid off when he secured a rank, earning him a place at the prestigious IIT Delhi in 2004. Here, a new world opened up for him, amplifying his experiences from Netarhat. The competitive environment and diverse opportunities pushed him to excel in both academics and extracurricular activities. He participated in internships abroad, played hockey, directed dramas, engaged in debates, and even delved into campus politics – embodying the lessons of seizing opportunities and pushing oneself to grow.

Shashank's well-rounded approach led him to secure a job as a consultant at a startup, setting him apart from his peers. While many students were hesitant to join a startup, perceiving it as risky, Shashank was captivated by the founders' passion during their pre-placement talks. He viewed the opportunity to become the startup's first employee as a unique and valuable experience, believing that being part of the company's history from its inception would be a significant achievement, regardless of his tenure.

The startup environment provided Shashank with a wide range of responsibilities, autonomy, and exposure to multiple projects. Working with renowned companies like Unilever, Nestlé, Britannia, PepsiCo, and Mother Dairy, he honed his professional skills in strategy, visualization, and entrepreneurial thinking. For nearly two and a half years, he worked closely with three first-generation entrepreneurs, gaining invaluable insights into the life of an entrepreneur, particularly during the 2008 recession.

As Shashank navigated his professional journey, he found himself asking some key questions. Throughout his life, he had been guided by the notion that the achievement of certain milestones would lead to a settled life. From his early days in the Bihar village, where someone advised him to attend Netarhat school, to his time at Netarhat, where he was told that scoring well in tenth grade would secure his future, Shashank had followed paths laid out by others.

This pattern continued even after he joined IIT, with people suggesting that securing a good job would be the key to a settled life. He worked diligently to land a prestigious consulting role. However, as he progressed in his career, he realized that the goalpost kept shifting. After two and a half years in consulting, he was advised to pursue an MBA from a renowned institution like Stanford or Harvard, with the promise that it would finally settle his life.

It was at this point that Shashank had an epiphany. He understood that life would never be truly settled and that he needed to take control of his own path. Instead of relying on others to dictate his next steps, he decided to introspect and identify what truly mattered to him.

Reflecting on his strengths, passions, and moments of genuine fulfilment, he realized that he derived the greatest satisfaction when he was involved with people on a large scale. Whether addressing a huge crowd in college, helping others excel in their tasks, or motivating people to perform better, he found joy in making a positive impact on others' lives.

This realization led Shashank to conclude that his true strength lay in his interpersonal skills. His ability to connect with people, understand their needs, and inspire them to achieve their goals was a valuable asset that he wanted to leverage in his future endeavours.

As he contemplated his next move, he found himself at the cusp of the startup revolution in India. Between 2008 and 2010, the entrepreneurial landscape was rapidly evolving. Books like Rashmi Bansal's *Stay Hungry, Stay Foolish*, Kishore Biyani's *It Happened in India* shed light on successful startup stories, and companies like Flipkart were making waves in the industry. The founders of Flipkart, also from IIT Delhi, had left lucrative jobs at Amazon to start their own venture, showcasing the potential of the startup path.

Immersing himself in this entrepreneurial world, Shashank discovered that starting his own venture would be an exciting alternative to the conventional MBA route. He realized that through entrepreneurship, he could touch the lives of thousands and make a meaningful impact on society. This revelation shifted his focus from personal gain to creating something of value for others.

He began exploring various sectors where he could make a difference, such as healthcare, education, and agriculture. His ideas revolved around working with the masses and creating scalable solutions to improve people's lives. As he delved deeper into these possibilities, a sudden thought struck him: why not focus on agriculture?

Agriculture had been an integral part of Shashank's formative years in the village. If those early experiences had played a crucial role in shaping his success, perhaps agriculture could provide the scale and impact he sought in his entrepreneurial journey. Excited by this prospect, he conducted extensive research to understand the potential and numbers associated with the agricultural sector.

The findings were astounding. Agriculture contributed 15 per cent to India's GDP, a fact that had never been discussed in his household. The total addressable market was in the billions of dollars, presenting an enormous opportunity for innovation and growth. Shashank realized that his unique exposure to both ends of the food chain – the farmer's perspective from his childhood and the consumer's side through his consulting work – positioned him well to tackle the challenges in this sector.

Armed with this new-found clarity, the aspiring entrepreneur realized that agriculture could be his calling. However, coming from a lower-middle-class family, job security remained a critical consideration. He decided to take a year to prepare himself mentally and financially for the leap into entrepreneurship while still employed at his consulting job.

During this preparatory period, Shashank immersed himself in the agricultural ecosystem. He leveraged his IIT network, connecting with successful entrepreneurs like Sanjeev Bikhchandani and Narayan Murthy to gain insights and guidance. He traversed the length and breadth of Bihar, visiting universities, farmers, wholesale bazaars, and mandis to gain a first-hand understanding of the market dynamics.

In this journey, he reconnected with his close friend Manish, a year his junior at IIT, who had always shown an inclination towards rural and social issues. Sensing a potential partnership, Shashank shared his plan to start an agricultural venture in Bihar with him. Despite being in the final year of his integrated course at IIT and in the middle of the placement process, Manish was eager to abandon the conventional path and join his friend. This commitment reinforced Shashank's belief that his idea was not just a dream but a vision that resonated with someone equally capable and passionate.

Over the next few months, the two friends immersed themselves in extensive research and preparation. They delved deep into the agricultural ecosystem, uncovering a recurring theme: a significant disconnect between farmers' needs and available solutions. This gap not only validated their mission but also highlighted the immense opportunity that lay before them. However, with each conversation, a challenging question surfaced: 'How will you do it?' It was a query that Shashank and Manish were eager to answer, fuelled by their growing confidence in the vast market potential.

After their thorough research, the next step was to find the ideal location to launch their venture. Their search led them to Vaishali, a district in Bihar that held both practical and symbolic significance.

Vaishali was not chosen randomly. It was a place steeped in historical and spiritual importance, being the first place that Gautama Buddha visited after his enlightenment. Situated midway between Gaya and Sarnath, it was where the Buddha had prepared himself for his further journey. This district was also the birthplace of Mahavir, the last Tirthankara of Jainism. This rich history imbued the place with a positive energy that resonated with the aspirations of the young entrepreneurs.

Practical considerations also played a role in their decision. Vaishali was Manish's native place and would provide them with a network of contacts to start with. For Shashank, it offered neutral ground, away from his own village where his family might have discouraged his agricultural pursuits. Strategically, Vaishali's proximity to Patna, the state capital, and the state agriculture university was advantageous. Its location in the Gangetic belt with its mixed crop pattern made it an ideal test bed for their venture.

Throughout 2009, the two friends made frequent trips to Vaishali, pitching their vision to the agricultural community and leveraging Manish's local connections. Their interactions allowed them to connect their research findings to real-world scenarios, gaining deeper insights into the multifaceted challenges faced by those working the land.

The struggles of the farmers there were complex and interconnected. They grappled with sourcing quality seeds, seeking reliable advisory services, and finding fair markets for their produce. Each step of the farming process was fraught with inefficiencies and hardships. The young entrepreneurs saw an opportunity to streamline this fragmented system. They envisioned a comprehensive solution: a seed-to-market package that would address farmers' pain points at every stage of the agricultural cycle.

Their proposal was revolutionary in its simplicity and scope. They would advise farmers on the most suitable crops they could grow, based on soil conditions, return on investment, and market demand. They would provide high-quality seeds and ongoing crop advisory services. Finally, they would handle the sale of the farmers' produce, ensuring fair prices for them. This holistic approach aimed to transform the entire agricultural value chain.

The response from the agricultural community was overwhelmingly positive. After numerous conversations, Shashank and Manish found a group of farmers willing to subscribe to their services across 250 acres of land. This initial success exceeded their expectations and filled them with optimism. They wondered why such an apparently simple solution hadn't been implemented before, and the ease with which they had convinced the farmers led them to believe they had stumbled upon a game-changing idea.

Buoyed by this positive reception, Shashank made the bold decision to resign from his job. Both he and Manish packed their bags and relocated to Vaishali in September, just as the sowing season was approaching in October. They were eager to transform their vision into reality, ready to hit the ground running and make a tangible difference in the lives of villagers.

However, reality soon dealt them a sobering blow. When they approached the farmers who had earlier shown interest, they were met with unexpected scepticism. 'You guys are really here?' the farmers asked, their disbelief palpable. This response shook the young entrepreneurs to the core.

The farmers explained that they were accustomed to city dwellers coming for surveys or research, making grand promises, and then disappearing without a trace. They hadn't expected the two young men to actually follow through on their words.

This revelation shattered the young men's initial optimism. The farmers who had initially agreed to participate now backed away, unwilling to entrust their lands to these young graduates from the city. It was a harsh lesson in the deep-rooted mistrust that existed between farmers and outsiders promising change.

Lesser individuals might have given up in the face of such a setback, but Shashank and Manish were made of sterner stuff. They persevered, continuing to engage with the farmers, addressing their concerns, and slowly trying to rebuild their trust. Their persistence paid off, albeit on a much smaller scale than they had initially hoped.

Instead of the 250 acres they had anticipated working with, they were given a mere three acres for their pilot project. Even this small plot was a collective contribution from fourteen farmers, who agreed more out of resignation to the persistence of the two young men than from genuine belief in their project. It was a humbling start, but the young entrepreneurs embraced it with enthusiasm. Working with just three acres of land contributed by fourteen sceptical farmers, they set out to prove their concept in a remote village of Vaishali.

Life during this period was both challenging and exhilarating. The remote villages they worked in had no hotels, and their limited funds meant such accommodations were out of the question. Instead, they relied entirely on the generosity of the local community. The villagers allowed them to use the community school building after 5 p.m., where each night they would join school benches together to create makeshift beds.

Their days were filled with non-stop activity. Rising with the sun, they would set out on their bikes, often covering 200 kilometres a day across rough rural roads, meeting farmers and organizing awareness programmes. With no restaurants or cafes in these remote areas, they frequently relied on the hospitality of farmers for their meals. In the evenings, they would return to their temporary shelter in the school, working late into the night on their laptops by the light of a single bulb, planning and strategizing for the next day.

Despite the hardships, these days were infused with a sense of joy and purpose. Every interaction with the farmers, every small intervention that brought about positive change, served as a powerful motivator. They were witnessing first-hand the impact of their work, which fuelled their determination to succeed.

In their pilot project, they introduced rajma (kidney beans) as an alternative crop, a novelty for the region. They provided comprehensive advisory services throughout the growing season, from sowing to harvest. Despite initial challenges, the crop yielded a reasonable harvest, and more crucially, the farmers secured good prices for their produce.

This small-scale success marked a turning point. It validated their concept and began to erode the farmers' initial scepticism. The sight of two IIT graduates living among the villagers, adapting to their lifestyle and delivering tangible results, created a buzz in the area. Their hands-on approach and commitment to working alongside the farmers earned them respect and trust within the community.

Buoyed by this success, Shashank and Manish expanded their efforts throughout 2011 and 2012. They moved from village to village with a straightforward pitch: 'Allocate a small portion of your land for a new crop. We'll provide the seeds, advisory services, and market access. You provide the land and labour.' This approach began to gain traction but came with its own set of challenges.

It was during this period of intense engagement that the team's composition began to evolve. In mid-2011, Amreen, a computer engineer from NIT Jamshedpur, reached out to them. Despite having lucrative job offers, he had chosen to return to his village in Bihar to work with the farmers there. After learning about their venture from local newspapers, Amreen expressed his desire to join their mission, seeing an alignment in their visions for rural development.

Shashank, recognizing the potential value of Amreen's technical expertise, welcomed him aboard. Privately, he was particularly excited about Amreen's background, knowing they would eventually need someone to build their tech stack and digitize their operations. Amreen relocated to Vaishali, joining Shashank and Manish in their modest life among the farmers.

Over the next few days, the team continued to grow as Abhishek and Adarsh joined in quick succession. Around the same time, Shyam, who had attended the same school as Shashank in Jharkhand and later studied at IIT and IIM, also came on board.

By early 2013, the group had grown to six individuals who shared a common passion and drive. This collective of professionals, each bringing unique skills and perspectives, proved to be a turning point for the venture. Shashank often reflects on this period as crucial in shaping the future of their enterprise, as the expanded crew allowed for more comprehensive problem-solving and innovation.

With this diverse group in place, the team's focus shifted to refining their approach and business model. They centred their efforts on sustainably improving farmers' income. Initially, they approached the problem analytically, thinking from their own perspective rather than the farmers'. They questioned the traditional farming practices, wondering why farmers stuck to certain crops when alternatives might offer better returns on investment (ROI) with less water consumption and labour.

This analytical approach led them to explore unconventional crop options, from papaya to baby corn, always seeking more remunerative alternatives for the farmers. However, on the ground, this innovative strategy revealed significant challenges.

The team realized that asking farmers to switch crops wasn't just a business proposition; it was a request to gamble with their livelihoods. These were families who had tilled the same lands, growing the same crops for generations. Trusting their entire year's income to the advice of young graduates, however well-intentioned, was a daunting prospect for the farmers.

A pivotal incident in 2012 highlighted this trust barrier. The team had convinced a farmer to grow baby corn, a crop with promising ROI. On the eve of harvest, when Shashank informed the farmer that it was time to collect the crop, the farmer hesitated. Having only worked with traditional corn varieties, the idea of harvesting what seemed to be a premature crop filled him with anxiety.

The following morning, as vehicles arrived for collection, the farmer called Shashank at 4 a.m., his voice laden with worry. 'Are you certain about this harvest? Have you verified everything? If this crop fails, I stand to lose my entire investment.' This incident starkly illustrated the enormous trust barrier they needed to overcome.

For these agricultural families, a failed harvest could mean financial ruin, inability to pay debts, or even loss of land. In contrast, Shashank and his crew, while passionate about their mission, could ultimately walk away if things didn't work out. This imbalance of risk was not lost on the farmers.

These experiences prompted a re-evaluation of their strategy. The team understood that building the necessary trust for introducing new crops would take years of consistent support, reliable advice, and proven results. This insight led to a significant pivot in 2013. Instead of focusing on new crops, they decided to become crop-agnostic. Their new pitch was simple yet powerful:

'Whatever you're growing, let us help you with inputs, advisory services, and selling your produce.'

This approach aimed to improve farmers' income from existing crops, reasoning that even with traditional staples like wheat or rice, there was scope to add value. While it might not double or triple incomes immediately, it could boost efficiency and earnings by 50–60 per cent through reduced input costs, better advisory services, and improved market access.

This strategy required less dramatic changes to be made by the farmers, making it easier to gain their trust. The team believed that after successfully managing a few cropping seasons and demonstrating success, they could then consider suggesting higher-ROI alternatives.

This crop-agnostic model became the foundation of what would eventually be known as the DeHaat approach. It marked a pivotal juncture in their journey, transitioning them to a scalable solution that addressed immediate needs while building trust for future innovations.

With this refined strategy in place, Shashank and his crew established their first agricultural services centre in 2013. Initially called the Farmers and Farm Foundation Centre, it served as a one-stop-shop for farmers, offering inputs, advisory services, training, and market access. The concept quickly gained traction, and within months, they opened a second centre in Vaishali district.

Within ten months, over 1,200 farmers had associated themselves with these centres, expressing satisfaction with the services provided. This rapid growth was encouraging. However, a new challenge emerged – identity.

The formal names 'Green Agro Revolution Limited' and 'Farmers and Farm Foundation' written on centre boards proved difficult for farmers to remember. Instead, farmers began referring to the group based on their initiatives: 'Papitawala' (papaya guys), 'IITwala' (IIT guys), 'Rajmawala' (kidney bean guys), or '*Baby corn wala aa gaya*' (the baby corn guy has arrived). These different names by which their venture was known were creating confusion and hindering brand recognition.

Recognizing the need for a unified, catchy, and relatable name, the team brainstormed. They wanted something easy to remember, something that resonated with their mission. Eventually, they settled on 'DeHaat', meaning 'village' in Hindi. The name not only reflected their focus on rural communities but also aligned with their vision of building a network or community of farmers.

The choice of DeHaat carried deeper significance. The group were acutely aware that agriculture was often perceived as unglamorous, even by farmers themselves. Many farmers felt shy about their profession, a sentiment that deeply troubled Shashank. He often thought, 'We IITians are here for them, working to improve their lives. Yet they themselves don't feel proud. The very term "dehaati" [villager] is used as a pejorative.' This realization fuelled their determination to contribute to changing this perception as their organization grew. They wanted to instil pride in the local people that they were from a 'Dehaat', emphasizing that it represented their culture and heritage.

As 2014 dawned, Shashank and his team set their sights on scaling their operations beyond the 30 villages in Vaishali district where they had perfected their model. Their research led them to adopt a micro-entrepreneur model, similar to Unilever's Shakti programme in FMCG and various microfinance initiatives. However, DeHaat's approach was unique – they provided year-round support to farmers, ensuring consistent activities and cash flow for their micro-entrepreneurs.

This insight led to the conceptualization of the rural agent-led business model. Armed with meticulously developed standard operating procedures (SOPs) and training guidelines, the team rapidly expanded from two to twenty DeHaat centres by the end of 2014. As they grew, a natural hub-and-spoke structure emerged, with the original centre serving as a hub for the new satellite centres.

Recognizing the potential of this structure, Shashank and his crew began envisioning a future network of multiple hubs, each connected to several DeHaat centres. This sparked a series of intense evening brainstorming sessions that would prove crucial to their future growth. After long days in the field, the team would huddle together, challenging themselves with ambitious scenarios: 'If we had unlimited resources, how would we expand? Where should our next hubs be located?'

These weren't mere daydreams but exercises in strategic foresight. The team mapped out potential locations, devised strategies for new districts, developed SOPs, and projected how their operations might evolve with scale. Their discussions touched on areas ranging from granular operational details to sweeping long-term goals, including the crafting of a comprehensive roadmap for the next three to five years and beyond. This forward-thinking approach, combined with their hands-on experience made for a unique blend of ambitious dreaming and pragmatic planning.

Despite their strategic planning, securing adequate funding remained a significant hurdle as the company scaled. Shashank and his crew demonstrated remarkable resourcefulness, employing a mix of strategies to keep operations afloat. They took on debt, borrowed from friends and family, and participated in competitions and challenges. Their efforts bore fruit as they won several prestigious awards including the first prize at 'Eureka'12' in the social category B-plan competition at IIT Bombay and the 'Mobile for Good' award from the Vodafone Foundation in 2015; they were also runners-up at the 'TATA Social Enterprise Challenge' at IIM Calcutta in 2014.

DeHaat's innovative approach garnered significant recognition, being featured as YourStory's 'Startup of the Month' and earning Shashank a place in Forbes 30 under 30 in 2014. They were also named an 'Amazing Indian' by Times Now in 2014 and secured the second Best AgriTech Startup of Asia title during the Rabo SustainableAg Challenge in Singapore. These accolades not only provided vital capital infusions but also validated their business model and mission.

Throughout this period, DeHaat operated as a for-profit agribusiness. Their revenue model involved charging a small margin on the agricultural inputs they supplied and on the produce they helped farmers sell, ensuring sustainability while keeping services affordable for farmers. This approach allowed them to reinvest in the company's growth while delivering value to their farmer clients.

Their perseverance and innovative approach caught the attention of NABARD (National Bank for Agriculture and Rural Development), which recognized the potential of these IIT-qualified individuals dedicated to rural development. NABARD's support, through grants and soft loans, provided crucial financial backing at a critical juncture, validating DeHaat's mission and enabling its continued growth.

By 2016, DeHaat was poised for further expansion, with plans for new hubs already mapped out from their earlier strategy sessions. Over the next year, they grew to four hubs (which they called nodes) within Bihar, connected to eighty DeHaat centres. An unexpected development played a pivotal role in this rapid expansion from twenty to eighty DeHaat centres. In early 2016, a small agricultural retailer approached Shashank with an unusual request: he wanted to become a DeHaat centre himself. Surprised, Shashank asked, 'Aren't you our competitor?'

The retailer's response was enlightening: 'Not at all, sir. As a small retailer, my business is seasonal and inconsistent. I've worked with many distributors who lack transparency and reliability. What DeHaat offers is different – a streamlined supply chain, year-round business opportunities, and farmer advisory services. This model allows me to focus on my strengths while benefitting from your institutional support and farmer relationships. It's a win-win situation.'

This interaction marked a turning point for the startup. They began incorporating small, conventional agri-stakeholders – input retailers and aggregators – into their network. These experienced agricultural businesspeople brought valuable industry knowledge, required minimal training, and were passionate about the sector. Their existing relationships and understanding of local dynamics proved invaluable to the company's strategy.

By leveraging these local partnerships, DeHaat's performance surpassed initial projections. Integrating existing agricultural businesses into centres provided instant credibility in new areas, enabling faster penetration into farming communities. This approach enhanced both the quality and reach of services, setting a strong foundation for DeHaat's continued growth in the agricultural sector.

Building on this momentum, from 2016 to 2018, Shashank and his crew focused on three strategic initiatives. First, they developed comprehensive crop advisory content for a wide range of crops, regardless of whether they were currently dealing with them. They collaborated with agricultural universities and leveraged their IIT networks to create this content. Importantly, they also began the process of digitizing this information, laying the groundwork for future technological advancements.

Second, they crafted a long-term geographical expansion strategy. Despite having only four hubs and an annual revenue of ₹30 crore (a significant achievement for a bootstrapped company), they ambitiously mapped out their future presence across India. They identified 296 districts that contributed to 55 per cent of India's agriculture, creating a detailed roadmap for expansion.

Last, they developed a technology roadmap. Since 2014, they had been creating basic app for farmers and CRM systems to provide them with advisory services. Now, they developed a comprehensive technology roadmap, anticipating future needs and capabilities that would support their scaling efforts.

However, this period of growth and planning was not without its challenges. In 2015–16, Shashank faced a significant personal and professional setback when Manish, one of the original co-founders, decided to leave the venture. It was a tough moment for him, as he had expected the core group to remain together for the long haul. The news took time to digest, marking a critical juncture in Shashank's entrepreneurial journey.

As 2018 approached, the company found itself in a rapidly changing landscape. The Indian startup ecosystem had experienced a boom, with companies like Ola, OYO, and Lenskart attracting substantial investments. However, a market downturn led investors to shift focus towards more stable sectors, including agriculture. This shift coincided with the rapid penetration of internet services into rural areas, creating a perfect storm of opportunity for agri-tech companies like DeHaat.

Shashank and his team recognized this pivotal moment. In earlier years, they had faced numerous rejections from investors who viewed agriculture as un-investable because of its perceived complexities and low-tech nature. Now, these same investors were showing interest. The perception of agriculture had shifted from that of a traditional, low-returns sector to a potential hotbed for innovation and technology-driven growth.

DeHaat, with its proven model of integrating technology with on-ground operations, was perfectly positioned to capitalize on this change. Shashank's years of persistence and the company's demonstrated success in Bihar had prepared them for this moment. They had a scalable model, a deep understanding of farmers' needs, and a vision for transforming Indian agriculture that now aligned with investor interests.

This convergence of factors culminated in a watershed moment for DeHaat in April 2019, when the company raised its first significant round of funding – $4 million from a consortium of investors. The impact was immediate and dramatic. DeHaat experienced 3x growth, with revenue jumping from ₹46 crore in March 2019 to ₹126 crore in March 2020. This rapid growth wasn't just about the influx of capital; it was a testament to the solid foundation the team had built over the previous years.

The company's full-stack approach, which reduced risks associated with seasonality and crop dependency, coupled with demonstrated execution skills, caught the attention of more investors. In just two and a half years, DeHaat secured a total of $220 million in investments – a stark contrast to its first seven years without external funding.

This substantial influx of capital, along with increased digital adoption during the COVID-19 pandemic, fuelled explosive growth. From April 2020 to the end of 2022, DeHaat expanded from eighty centres and four hubs to an impressive 10,000 centres and 106 hubs. Revenue soared from ₹126 crore in March 2020 to an astounding ₹1,500 crore two years later. The company's footprint expanded to 125 districts across India, and its employee base grew from 400 to 2,400, marking DeHaat's transformation from a regional player to a national agricultural powerhouse. Their meticulous planning, roadmaps, SOPs, and guidelines developed during those late-night strategy sessions were now bearing fruit.

Behind this remarkable growth also lies a story of organizational culture and strategic foresight cultivated from the very beginning. The organizational culture at DeHaat was shaped by the original team of six who worked closely together. They engaged in heated debates but always united behind decisions once made. This approach fostered inclusivity, transparency, and mutual care, which proved invaluable as the company scaled. The versatility of the founding group was another crucial factor. In the initial seven years, each member wore multiple hats, performing roles from HR to finance, from managing inputs to outputs. Shashank often likened this to a cricket fielding strategy, where everyone runs to where the ball goes.

This comprehensive understanding of the entire system proved crucial when scaling. As the company grew, each founding member took charge of a specific domain, allowing for focused expertise while maintaining a holistic view of the enterprise. This approach enabled informed decisions across all aspects of the business, ensuring balanced growth and quick problem-solving.

Equally important was the company's strategic foresight. While many were shocked by its rapid rise from ₹46 crore in 2019 to ₹1,500 crore in revenues by 2022, they didn't see the initial eight years of hard work and preparation. Shashank often uses the analogy of a bamboo plantation: what people see is the rapid upward growth, but they don't see the deep, extensive root system developed over many years.

This principle of laying strong foundations before visible growth is evident in various aspects of DeHaat's business. For instance, they began preparing for the shift towards biological inputs in agriculture years before it became a global trend. This foresight has paid off, with their standalone business in biologicals – which includes organic fertilizers, biopesticides, and biostimulants – set to exceed ₹200 crore this year.

Another example is their venture into exports. Though this business began only two years ago, it has already reached ₹200 crore in turnover. This rapid growth is the result of meticulous preparation. Years before entering the export market, the team had anticipated growing international demand for quality Indian produce. They prepared by improving crop quality, implementing stringent quality-control measures, and building relationships with international buyers.

Despite its rapid growth, DeHaat has remained true to its core principles – staying grounded, maintaining close relationships with farmers, believing in technology, and thinking ahead of the curve. As Shashank often says, 'As an individual, if you have a long-term perspective, things will fall into place.' This philosophy has guided the company's approach to growth and innovation. They have a clear vision and goals but aren't bound by rigid timelines, believing that with the right fundamentals in place, they will eventually reach their objectives.

Today, DeHaat stands as a category leader in the agri-tech space, having successfully addressed initial scepticism about their model's scalability and agriculture's profitability. With a clear path to crossing ₹7,000 crore in revenue and plans for a future market listing, the company has proven that it's possible to build a successful, scalable, and profitable business in the agricultural sector.

As DeHaat continues to grow, Shashank often reflects on the journey from those early days in Vaishali. The barefoot boy who once navigated the rugged terrain of rural poverty has indeed left an indelible mark on the fields he sought to transform. From sleeping on school benches to leading a company valued at over $700 million, Shashank's story embodies the transformative power of resilience, determination, and unwavering belief in the value of agriculture.

Today, DeHaat stands at the forefront of India's agri-tech revolution, serving over 2 million farmers across the country. Yet, for Shashank and his team, this is just the beginning. They envision a future where technology and tradition blend seamlessly to empower every Indian farmer, making 'DeHaati' a badge of honour rather than a term of derision.

As India grapples with the challenges of feeding a growing population in the face of climate change, companies like DeHaat are poised to play a crucial role. By continuing to innovate and adapt, they aim to not only improve farmers' livelihoods but also ensure food security for the nation.

# 10
# Sowing the Seeds of Change: Lessons from the Field

### Techies Who Talk to Plants: Entrepreneurs Transforming the Future of Agriculture

Our journey through this narrative has been nothing short of inspirational. We've witnessed how a group of visionary individuals, many with backgrounds in technology and engineering, chose to apply their skills and innovative mindsets to one of the world's oldest and most vital industries: agriculture.

These entrepreneurs didn't just apply technology blindly to agriculture; they listened to farmers, immersed themselves in rural realities, and leveraged cutting-edge technology to address age-old agricultural challenges. Their stories exemplify a powerful fusion of technological expertise and agricultural wisdom, demonstrating how innovation can flourish when silicon meets soil.

From BigHaat's data-driven approach to WayCool's supply chain revolution, from SatSure's satellite-powered insights to Aquaconnect's tech-enabled aquaculture solutions, and from Eggoz's quality-focused poultry farming to DeHaat's comprehensive farmer services platform, each venture showcases a unique approach to agricultural transformation. These entrepreneurs, along with the visionaries behind S4S Technologies and Prompt Dairy Tech, have not just built businesses; they've sown the seeds of a new agricultural paradigm in India.

As we reflect on their journeys, several key lessons and themes characterizing this new wave of agri-tech innovation emerge:
1. **Immersion in the farmer's world:** Perhaps the most critical lesson is the importance of a deep, first-hand understanding of the Indian farmer's reality. Almost universally, these entrepreneurs spent significant time living among farmers, experiencing their daily challenges, and understanding their fears and anxieties. This immersion was not a brief market research exercise but often extended for months or even years. It formed the bedrock of their business models and drove their innovation. As Steve Jobs famously said, 'You've got to start with the customer experience and work backwards for the technology. You can't start with the technology and try to figure out where you're going to try to sell it.' These agri-tech pioneers embodied this principle, starting not with technology but with a deep understanding of farmers' needs.
2. **The power of purpose:** Each of these entrepreneurs was driven by a mission larger than themselves. Whether it was improving farmers' livelihoods, enhancing food security, or promoting sustainable agriculture, their ventures were founded on a strong sense of purpose. This purpose-driven approach not only fuelled their perseverance through challenges but also inspired others to join their cause.
3. **Technology as a means, not an end:** While these are stories of 'agri-tech', it's striking that most of these entrepreneurs didn't start with technology. Instead, they began with a problem they observed in the agricultural sector and then sought or developed technologies to address it. This approach ensured that their solutions were relevant, accessible, and truly beneficial to farmers.

4. **Importance of bootstrapping and resilience:** Several of these entrepreneurs, like those at DeHaat and Prompt Dairy Tech, spent years bootstrapping before securing significant funding. This period of resource constraint forced them to be innovative, efficient, and deeply connected to their customers. It built resilience, which served them well as they scaled.
5. **Innovative funding in a capital-scarce environment:** Faced with the reality that their primary customers – farmers – often couldn't pay much, these innovators had to be exceptionally creative in generating funds. They worked frugally, lived frugally, and explored every avenue for generating capital – from participating in competitions and awards to leveraging government schemes. This resourcefulness proved that lack of venture funding need not be an insurmountable barrier to building a successful business.
6. **Holistic approach to problem-solving:** The most successful ventures didn't just offer point solutions but took a holistic approach to addressing farmers' challenges. From providing inputs and advisory services to facilitating market linkages, they created comprehensive ecosystems that added value at multiple points in the agricultural value chain.
7. **Thorough preparation and diverse skill sets:** Many spent at least a year researching the market, understanding the potential for scalability, and developing critical business skills before launching their ventures. Moreover, these entrepreneurs weren't specialists in a single field. They developed knowledge across various domains – from agriculture and technology to finance and marketing. This generalist approach, combined with hands-on learning, enabled them to navigate the complex, multifaceted challenges of the agricultural sector.

8. **Empowering teams through autonomy:** A key factor in the rapid scaling of many of these businesses was the autonomy granted to team members. This approach not only accelerated growth but also built robust, adaptable organizations capable of seizing new opportunities.
9. **The role of mentorship and ecosystem support:** The influence of mentors like Hemendra Mathur and the support of various ecosystem players have been crucial in nurturing these ventures. This underscores the importance of building a robust support system for agri-tech startups in India.
10. **Scalability and profitability in agriculture:** These stories challenge the notion that agriculture cannot be a scalable, profitable business. By leveraging technology, data, and innovative business models, these entrepreneurs have shown that it's possible to build large, impactful, and financially viable agri-tech ventures.

As we look to the future, the potential of agri-tech in India seems boundless. With a large agricultural base, increasing smartphone usage, and a supportive policy environment, India is poised to become a global leader in agricultural innovation. The entrepreneurs featured in this book are just the vanguard of a larger movement that is reimagining Indian agriculture for the twenty-first century.

However, challenges remain. Climate change poses an existential threat to traditional agricultural practices. The digital divide still excludes many farmers from accessing technological solutions. And the complex, fragmented nature of Indian agriculture continues to present hurdles for scalability.

Yet, if the stories in this book teach us anything, it's that with innovation, perseverance, and a deep commitment to farmers' welfare these challenges can be overcome. As these agri-tech ventures continue to grow and new ones emerge, they have the potential not only to transform Indian agriculture but also to provide solutions for global food security and sustainability.

*Techies Who Talk to Plants* is more than just a catchy title – it's a powerful metaphor for the bridge these entrepreneurs have built between the world of technology and the realm of agriculture. By learning to 'talk to plants' – to understand deeply the needs of crops, farmers, and the entire agricultural ecosystem – these techies have opened up new possibilities for sustainable, efficient, and profitable farming.

In the fertile fields of India, a new chapter in the story of agriculture is being written. The seeds of change have been sown by these visionary entrepreneurs, and they're already taking root. As we look to the horizon, we can almost see the first green shoots of a revolution emerging – a revolution that promises to feed millions, empower farmers, and cultivate a more sustainable future for us all.

The harvest of their labour may not be fully realized for years to come but make no mistake – it will be bountiful beyond our wildest dreams. And when that day comes, we'll remember these techies who dared to talk to plants, and in doing so, changed the very fabric of Indian agriculture. Their legacy will not just be written in the annals of business history but will also be remembered by the prosperous rural communities and well-fed families across India and beyond.

# Acknowledgements

This book would not have been possible without the extraordinary entrepreneurs who generously shared their stories, struggles, and triumphs with me. To Sachin Nandwana, Prateep Basu, Karthik Jayaraman, Rajamanohar, Hemendra Mathur, Abhishek Negi, Nidhi Pant, Shridhar Mehta, and Shashank Kumar – thank you for your time, candour, and willingness to let me chronicle your remarkable journeys. Your dedication to transforming India's agricultural landscape has been nothing short of inspirational.

I am deeply grateful to Harsh Mariwala for contributing the insightful foreword that perfectly frames the importance of innovation in agriculture. Your perspective adds tremendous value to this narrative.

A special thanks to Hemendra Mathur, who deserves particular recognition for his immense contribution to this book. Though he wasn't involved in its inception, once the book began taking shape, he became a constant source of inspiration and mentorship. His willingness to share knowledge, explain the intricacies of the agricultural sector, connect me with entrepreneurs, motivate me during challenging times, and persistently follow-up played a major role in bringing this book to publication. This work would not be what it is without his guidance and support.

I would also like to express my gratitude to Sanjiv Rangrass, former CEO of ITC Agri Business, for offering his encouraging support throughout this project. His belief in this book and his assistance in sharing it with others as well as collecting valuable feedback has been instrumental in bringing this work to a wider audience.

My heartfelt thanks to the team at Bloomsbury India, particularly Paul Kumar and the editors whose guidance, patience, and belief in this project transformed a collection of stories into a cohesive exploration of India's agri-tech revolution. Your editorial insights strengthened this book in countless ways.

This journey of discovery would not have been possible without the numerous farmers, field workers, agricultural scientists, and industry experts who shared their knowledge and perspectives. Your contributions have enriched my understanding and helped paint a more complete picture of the agricultural ecosystem.

To my family – my wife Shabna and my ten-year-old daughter Aaima – thank you for your unwavering support throughout the research and writing process. Your patience, encouragement, and love sustained me through this journey.

My colleague Mahmath A. at D-Cube Designs deserves a special mention for maintaining our creative consultancy while I pursued this passion project. Your professionalism and dedication allowed me the freedom to explore this new frontier.

Finally, to all the readers who will pick up this book – I hope these stories inspire you as much as they have inspired me. The future of Indian agriculture is being reimagined by these techies who dared to talk to plants, and I am privileged to have played a small part in telling their stories.

# Index

Achrekars, 140
agribusiness in India, 1
agricultural trading firms, 3
agricultural value chain, 2
agri-fintech solutions, 9. *See also* SatSure
agri-input market in India, 26
AgriStack, 150–151
   vision of, 151
agri-tech startups, 8–10, 12–13, 147, 254
   challenges, 10–11
   direct-to-farmer and direct-to-consumer models, 10
   effort and vision, 17–18
   impact, 15–16
   lessons and themes, 251–253
   lessons learnt, 16–17
   reality, 14–15
   skill development, 10
   sustainable practices, 10
   technology adoption, 10
agri-tech startups in India, 1
AgTech solutions, 66
AI-enabled technologies, 27
AI-powered crop monitoring, 15
Ali, Muhammad, 203
*Alien,* 52
Amazon, 40, 231
Amreen, 237–238
Amul, 5–6, 222
Aquaconnect, 113, 250
   AquaCRED programme, 129–130, 133
   Aqua Partner programme, 129, 131
   AquaSAT platform, 134–136
   business model, 122–124, 127–128
   challenges with retailers, 128–129
   farm input solutions and post-harvest linkages, 123–124
   geographical footprint, 133, 137–138
   identification and on-boarding of biotech companies, 131
   impact of, 131–132
   issue of access to working capital, 129
   on-boarding of seafood processors, 124
   overnight delivery service, 130
   R&D division under 'Dr. Grow,' 137
   retail network, 131–132
   for seafood buyers, 132–133

seafood value chain, 134–139
   as source of market intelligence, 131
   'Story of Shrimp' program, 135
   WhatsApp-based customer outreach, 130
   working of FarmMOJO app, 124–126
aquaculture, 118–120
   challenges in value chain, 121–122
   critical parameters, 120
   pre- and post-production stages, 121
*Armageddon,* 52
ARPU (Average Revenue Per User), 117
artificial intelligence and robotics, 12
Ashok Leyland, 85
Asimov, Isaac, 51
automated Milk Collection System, 212–213
   benefits of, 213
automobile industry, 87

Bajaj Auto, 82
Baring Private Equity, 74, 76
Barwale, Dr Badrinarayan, 2
Basu, Prateep, 71–77. *See also* SatSure
   academic journey, 51–55
   business idea, 62
   enthusiasm for space and astronomy, 51
   GSLV MK-III project and PSLV launches, 55–56
   at International Space University (ISU), France, 56
   job at ISRO, 55–56
   life at Bengaluru, 57
   research and consulting job, 57
   societal applications of satellite data, 58–60, 71–72
   story of driver's tale of hardship, 60–61
Beyond Next Ventures, 42
Bhabha Atomic Research Centre (BARC), 184
Bharat, 63
Bharat Innovation Fund, 73, 148
BigHaat, 19, 33, 250
   business model, 41–42, 49
   Crop Doctor app, 47
   deliveries, 36
   as a farmer-centric platform, 39–40
   financial reserves, 43
   focus on horticulture seeds, 33–34
   founders' vision for, 49–50
   funding for, 41–42
   influence in agri-inputs supply chain, 44
   Kisan Mudra, 47
   Kisan Vedhika, 47
   marketing strategy, 48
   missed-call system, 35
   omni-channel approach, 35
   partners, 42, 45, 48
   partnership with India Post, 36
   partnership with US Agriseeds, 37–38, 44

# Index

primary customers, 39
real-time alerts and advisories, 45–47
resilience and growth, 48
setbacks, 41
stakeholder management and technology development, 48
sustainability development programme (SDP), 46
trust and transparency, 45–46
Bikaner, 2
Bikhchandani, Sanjeev, 233
Bill & Melinda Gates Foundation, 65
biotechnology and genetic engineering, 11–12
blockchain technology, 12
bootstrapping, 110, 252
Britannia, 230

Carrefour, 142
Centre for Development of Advanced Computing (CDAC), 23–25
  DESD programme, 24–25
Chawla, Kalpana, 51
Clarke, Arthur C., 51
climate change impact, 11, 15, 135–136, 183, 249, 253
climate-tech startups, 153
Cochin University of Science and Technology (CUSAT), 229
contract farming, 4–5, 29
cooperative movement, 5–6
  challenges, 6
corporate agriculture, 3–5
  impact of, 4–5

credit cooperatives, 5
crop loan
  approval time for banks, 70–71
  CIBIL score, 70
  issues faced by farmers, 70–71
  satellite and AI-driven farm loan enablement, 71
cutting-edge projects, 25

Dasari, Vinod, 86–87
data-driven agriculture, 11
Datta, Arindom, 73
DeHaat, 240–241, 243, 250, 252
  centres, 241–244, 246
  core principles, 248
  end-to-end solutions and services, 244–245
  funding for, 242, 246
  growth, 248–249
  micro-entrepreneur model, 241
  NABARD's support, 243
  organizational culture, 247
  recognitions and awards, 242–243
  revenue, 246
  standalone business in biologicals, 248
  strategic initiatives, 244–245
dehydration, 189
  benefits of dehydrator, 190
  market opportunities for dehydrated products, 191–192
  prototype development, 189–190
Digital AgriStack, 151

digital divide, 10
Digital India, 8
e-commerce platforms, 40
egg market, 166, 178
    Bihar, 166–167
    challenges in supply chain, 167
    egg production and consumption, 167
Eggoz, 155, 182, 250
    branding, 176
    business model, 181
    challenges and opportunities, 178, 181
    core values, 180
    delivering eggs, 181
    development of brand, 175–176
    funding, 177–178
    geographical footprint, 177, 179
    impact of COVID-19 pandemic, 176, 179
    partnership with online platforms, 177
    pricing, 176
    state-of-the-art egg processing centres, 181
    strategic approach to hiring, 180–181
    value-added products, 179–180
emergency-light business, 21

Faheem, 74
family-owned enterprises, 1–3
    challenges, 3
    collaboration between agri-tech startups and, 11
    succession planning, 3
    types, 2–3
farm equipment manufacturers, 2
Farmers and Farm Foundation Centre, 240
farmers' needs, understanding of, 163–164, 187–189, 232, 251
farm management software, 9
'Fat'omatic,' 213–214
Federer, Roger, 203
financial crisis, 2008, 145
fintech, 67
Flipkart, 13, 40, 231
Food Corporation of India (FCI), 7
food preservation methods, 188–189
food processing units, 2
FOSS Analytical, 219
Freshey's, 99–100
future of agribusiness in India, 11–12

Gandhi Krishi Vigyan Kendra (GKVK), 30
Gates Foundation, 174
global and local food supply chains, 142
Godrej Agrovet, 4
Google, 26
government initiatives in agriculture, 5–8
Green Agro Revolution Limited, 240
Green Revolution, 2
Gujarat Cooperative Milk

Marketing Federation, 5–6
Haldiram, 2, 202
HDFC Bank, 76
Hexolabs, 115–118
   in African markets, 117
   business model, 118
   operator-driven value-added services (VAS), 118
holistic approach to problem-solving, 77, 137, 222, 235, 247, 252
HUL, 98
hypermarkets, 105–106

ICICI Bank, 68, 76
ICICI Foundation, 174
IIT Madras, 80–81
*Independence Day*, 51
Indian agri-tech ecosystem, 13
Indian Farmers Fertilizer Cooperative Limited (IFFCO), 6
Indian Institute of Space Science and Technology (IIST), 53–54
Indian Institutes of Management (IIMs), 9, 61, 65, 84–85, 141–142, 148, 238
Indian Institutes of Technology (IITs), 9, 53, 79–82, 84–85, 113–115, 141, 155–157, 224, 229, 238
Indian School of Business (ISB), 84
Indian Space Research Organization (ISRO), 53
India's agricultural landscape, 1
India's digital infrastructure, 40–41
input supply cooperatives, 5
International Finance Corporation (IFC), 151
IOT-based precision water control system, 29
ITC, 4, 98
*It Happened in India* (Kishore Biyani), 231

Jayaraman, Karthik, 111. *See also* WayCool
   academic journey, 79–81, 84
   aptitude for problem-solving, 79
   at Ashok Leyland, 85–87
   business and management skills, 85–86
   business idea, 90
   emerging leaders programme, 86
   fundraising and capital allocation, 105, 110
   leadership, 86
   logistics and supply chain management, 87–90, 100–101
   on market dynamics in India, 105
   at McKinsey, 84–85
   passion for mechanics, 78–79
   at Tata Motors, 82–83
   at Timken, 83–84
Jobs, Steve, 251
Jordan, Michael, 203

Kaizen concept, 28
Kalam, Dr A.P.J. Abdul, 53
Kisan Rail, 7

Korean BBQ food trucks, 87
Kumar, Shashank, 225–226
   academic journey, 227–231
   agricultural venture in Bihar, 233–237
   analogy of a bamboo plantation, 247
   childhood experiences, 225–228
   crop-agnostic model, 239–240
   on crop rotation, 238–239
   job as a consultant, 230
   professional journey, 230–231
   seed-to-market package, 234–235
   understanding of agriculture sector, 232
Kumar, Uttam, 162, 164, 170

Lays, 202
Lehman Brothers collapse, 145
Lenskart, 245
Liquid Propulsion Systems Centre (LPSC), 55
logistics and supply chain management, 87–90, 100–101. *See also* WayCool
   aligning with demand, 89–90

Maggi, 179
Mahindra, 81–82
Mahindra & Mahindra, 4
Mahyco (Maharashtra Hybrid Seeds Company), 2
Make in India, 63
Manish, 233–237, 245
Marico, 98
marketing cooperatives, 5
Maruti Suzuki, 81
Mathur, Hemendra, 17, 65, 73, 202–203, 253
   academic journey, 140–142
   analysis of performance of agri-tech startups, 152–153
   collaboration with Beanstalk, 152
   at College of Technology of Engineering in Udaipur (CTAE), 140–141
   concept of AgriStack, 150
   drafting of Food Processing Policy of India, 143–144
   embrace of entrepreneurship, 153–154
   on FICCI taskforce committee on agri-startups, 149
   financing of agri-tech ventures, 147–148
   KSA Technopak experience, 142–143
   launch of ThinkAg, 148–150
   love for market research, 142
   as a mentor, adviser, or board member, 150
   NIFTEM project, 144
   publications, 150
   at Rabobank's advisory and research division, 143
   role in National Horticulture Mission (NHM), 144–145
   at SEAF India Investment Advisers, 146–147

understanding of global and local food supply chains, 142
work on Digital Public Goods, 151
at YES Bank, 145
Matre, Siddhant, 159–160
McKinsey, 84
Mega Food Parks Scheme, 8
Mehra, Arul, 75
Mehta, Shridhar, 205. *See also* Prompt Dairy Tech
  academic performance, 206
  computer learning, 206–207
  development of automated Milk Collection System, 212–213
  development of software for inventory control and sales tracking, 207
  electronic weighing scale business, 208–211
  establishment of network of distributors and service centres, 215
  establishment of Zalak Computer Centre, 207, 209
  family's power tool business, 205–206
  milk collection process, understanding of, 209–211
  milk measurement process, 209–214
mentorship, 154, 253
Minimum Support Prices (MSP), 7
MIT, 81
mobile engagement and content consumption, 117
Moser, Dr Roger, 64
Mother Dairy, 230
Murthy, Narayan, 233

Naidu, Shri Ram Mohan, 64
Nandwana, Sachin, 33, 35, 37–39, 41, 43. *See also* BigHaat
  academic journey, 21, 23–24
  at Bharat Electronics Limited (BEL), 23–24
  development of agri-tech company, 32–33
  development of electronic product for Irish firm, 26
  early farming experiences, 19–20
  entrepreneurship and self-reliance principles, 20–21, 28–29
  at Honeywell, 25, 27, 32
  innovations, 21, 27
  at Instrumentation Limited, Kota, 21–23
  lessons about product-market fit and timing, 26
  problem-solving skills, 22–23
  quality-control functions and preliminary testing, improvements in, 22
National Agricultural Cooperative Marketing Federation of India (NAFED), 7
National Agriculture Market (eNAM), 7
National Agro Foundation, 92
National Bank for Agriculture and Rural Development

(NABARD), 7, 110, 243
National Horticulture Mission (NHM), 144–145
National Institute of Food Technology Entrepreneurship and Management (NIFTEM), 144
National Seeds Corporation (NSC), 7
Negi, Abhishek, 155
    academic journey, 156–157
    acquisition of poultry farm, 171–172
    building own poultry farm, 169–171
    building poultry supply chain, 175–176
    challenges faced by farming community, understanding of, 163–164
    childhood experiences, 155–156
    consumer-facing egg business model, 171
    egg production, understanding of, 168–169
    'farming-as-a-service' model, 165–166
    idea of one-way taxi fares, 159–160
    importance of human connection, 156
    maintenance of poultry farms, 173–174
    path of entrepreneurship, 159
    poultry supply chain in Bihar, understanding of, 166
    problem-solving skills, 158
    strategies to improve mobile network quality, 158
    at Vodafone India, 157–158
Nestle, 98, 222, 230
Nokia, 27, 114
Norwegian aquaculture industry, 123
Nukala, Sateesh, 25–26, 30, 33, 35, 37–39, 41, 43. *See also* BigHaat
    leadership trait, 49

Ola, 160–161, 245
organic farming, 29
OYO, 245

Pant, Nidhi, 183. *See also* S4S Technologies
    academic journey, 184–185
    childhood, 183–184
    participation in writing, project management, and team leadership, 185
    on plight of women farmers, 194–195
    professional journey with S4S Technologies, 204
    understanding of needs and challenges of rural communities, 187–189
Pardhasaradhy, Chinna, 95
Paytm, 13
PepsiCo, 230
Pepsi's tomato farming initiative, 4
Pizza Hut, 142
poultry industry in India, state

of, 167–168
precision farming, 9
*Predator,* 52
processing cooperatives, 5
Prompt Dairy Tech, 215–224, 250, 252
    B2B relationships, 222
    cloud-based system for dairy supply chain, 217–218, 222
    collaboration with Vodafone, 218
    contract with Gujarat's dairy industry, 218
    creation of 'pure milk fingerprint,' 219
    funding issues, 216
    growth and diversification, 220
    innovations at, 221–223
    IoT system for monitoring bulk milk containers, 219–220
    MilkoChill, 222
    MilkoScreen, 219
    workforce and products, 221
Proof of Concepts (PoCs), 67
Public Distribution System (PDS), 7
public-private partnerships (PPPs) in agribusiness, 7
public-sector enterprises, 5–8
Purdue University, 64, 81–82
purpose-driven approach, 251

Rabobank, 65, 69, 143
Rajput, Ashish, 159–160
Raju, Abhishek, 61–62, 71
Rashmit, 63, 71
Redbus, 13
regulatory environment, 11
Roder, 160
    challenges and closure, 161–162
    as digital cab aggregator for outstation travel, 160
    funding, 161
rural agent-led business model, 241

SaaS-based e-commerce platform, 38
Samunnati Finance, 69
SatSure, 62–63, 250. *See also* crop loan
    achievements and recognition, 68
    applications of satellite data, 71–72, 76–77
    approach in building customer credibility, 68
    challenges, 68
    customers, 68, 76–77
    direct financing to farmers, 70
    employees of, 65
    financial product in partnership with CIBIL, 77
    first commercial sale, 69
    first project, 64–65
    fundraising process, 66–67, 69, 72–76
    as a novel agri-tech firm, 63
    revenue concentration risks, 68, 70
    revenue stream, 75
    total addressable market (TAM), 74
    value proposition of, 75

scalability and profitability in agriculture, 11, 253
seafood industry
  exports, 119–120
  role in global food security, 120
  as a source of livelihood in rural and coastal communities, 120
seed and fertilizer companies, 2
Senthil, 92
Singh, Aditya, 162–163, 170
SMS-based search engine (SMS Wiki), 117
Somasundaram, Rajamanohar (Raj), 112–140. *See also* Aquaconnect
  ABC package, 117
  academic journey, 113
  on aquaculture, 118–120, 138–139
  development of mobile apps and games, 115–116
  edutainment concept, 115
  Internet banking solutions project, 113
  opportunities with mobile technology, 114–116
  on shrimp sales and prices, 118–119
  strategic approach to pricing and partnerships, 117
  study of Africa as a market, 117
Sonalika Group, 2
space technology entrepreneurs, 56
Sravan, 63–64
S4S Technologies, 183, 186, 250
  advantages of food processing equipment, 199
  aggregator role, 193–194, 203
  approach of profitability with social impact, 203
  as a dedicated agri-tech company, 192
  development of a direct-to-consumer (D2C) brand, 201–202
  effect of Mathur's guidance and extensive network, 202–203
  funding, 195–196, 203
  idea of food preservation at farmer level, 188–191
  impact of COVID-19 pandemic, 202
  market opportunities for dehydrated products, 191–192
  partnerships with stakeholders, 198
  recognition and accolades, 191
  sales challenges, 190
  support and guidance to women farmers, 196–201
  training programmes, 197
  transformation of agricultural value chain, 195
Startup India, 8, 63
*Stay Hungry, Stay Foolish* (Rashmi Bansal), 231
supply chain optimization, 9

Suresh, Dr B.N., 54
sustainable and climate-resilient farming, 11

Tata Motors, 78, 81
team autonomy, 82–83, 102, 230, 253
technology as a means, 251
*Terminator* series, 52
TESCO, 142
ThinkAg, 148–149
    as agri-food-fintech platform, 149
    challenges during pandemic, 149–150
Tidke, Vaibhav, 186
Timken, 83
trust building, 10–11

Uber, 160–161
Unilever, 230
Unilever's Shakti programme, 241
University Department of Chemical Technology (UDCT), 185–187
US Agriseeds, 37–38, 44

Vannum, Kiran, 42, 45
Velvette shampoo, 116
vertical and urban farming, 12
vertical integration, 4
Vikram Sarabhai Space Centre (VSSC), Trivandrum, 54
Vinay, 114–115

water scarcity in Himalayas, 184
WayCool, 78, 250
    acquisitions, 95

automated warehouse, 93
in Bengaluru, 94–95
business model, 90–91
Carrying and Forwarding Agents (CFAs) and distributors, 104
challenges, 91–92, 97, 110
collaborative problem-solving methods, 107
consumer-driven model, 101
distribution of FMCG products, 98
establishment of dal supply chain, 98–99
expansion of business, 91–93, 102
field-centric model of, 108
franchises, 101
Freshey's acquisition, 99–100
growth, 108
intra-state sourcing and delivery, 99–100
inventory management system, 101
Madhuram brand, 97
Marico products distribution, 98
marketing strategies, 102
milk and milk product sales, 99–100
philosophy of mutually beneficial relationships, 107–108
product diversification, 95–100
reverse logistics approach, 93
rice procurement and scale,

96–97
strategic retreats, 103
in Tindivanam, Tamil Nadu, 93
warehouses, 104
weather-risk forecasting, 63

Wells, H.G., 51
Williams, Serena, 203

Yahoo, 52
Yes Bank, 65

# About the Author

Shah M.M. is a distinguished Business Strategy and Design Thinking consultant whose journey from industrial designer to chronicler of agricultural entrepreneurship reflects a deep commitment to innovation and societal impact. With a master's degree in design from IIT Kanpur (2004) and over a decade of experience launching more than 300 products across various industries, Shah has always been at the forefront of user-centred design.

As the founder of D-Cube Designs, a leading design consultancy in Chennai, India, he has established himself as a trusted adviser to CEOs, startup founders, and aspiring entrepreneurs, providing valuable insights on branding, user-centred design, creativity techniques, and building competitive advantages.

A few years ago, Shah found himself captivated by the spirit of innovators who dared to dream big. This led him to share insights through his blog, which now includes over 600 articles on business strategy and related topics. Shah's growing interest in entrepreneurial stories led him to explore the world of agri-tech, where he discovered powerful narratives of hope and transformation.